Materials Science: Structure and Characterization of Materials

Materials Science: Structure and Characterization of Materials

Editor: Kevin Chapman

NY RESEARCH PRESS
New York

Published by NY Research Press
118-35 Queens Blvd., Suite 400,
Forest Hills, NY 11375, USA
www.nyresearchpress.com

Materials Science: Structure and Characterization of Materials
Edited by Kevin Chapman

International Standard Book Number: 978-1-63238-600-7 (Hardback)

Cataloging-in-Publication Data

Materials science : structure and characterization of materials / edited by Kevin Chapman.
p. cm.
Includes bibliographical references and index.
ISBN 978-1-63238-600-7
1. Materials science. 2. Materials. I. Chapman, Kevin.
TA403 .M38 2018
620.11--dc23

Printed in China.

Contents

Preface

The study of designing new and improved materials is known as materials science. It is an interdisciplinary field which incorporates the elements of chemistry, metallurgy, physics, engineering and mineralogy. The topics studied under this subject are crystallography, thermodynamics, bonding, structure, kinetics, etc. This book elucidates the concepts and innovative models around prospective developments with respect to materials science. The book is a valuable compilation of topics, ranging from the basic to the most complex theories and principles in this field. Those in search of information to further their knowledge will be greatly assisted by this textbook.

Given below is the chapter wise description of the book:

Chapter 1- The study, measure and analysis of a material's structure and properties is called as material characterization. Microscopic techniques such as the electron microscope and secondary ion mass spectrometry as well as macroscopic techniques like thermal analysis and density calculation are part of this process. This is an introductory chapter which will introduce briefly all the significant aspects of material characterization.

Chapter 2- X-rays are electromagnetic radiation with wavelength ranging from 0.01 to 10 nanometers. X-rays are used for structural analysis of materials. It is known to interact with matter through photoelectric absorption, Compton scattering, and Rayleigh scattering. The chapter also explores the concepts of Bragg's law, powder diffraction and crystallite among few others. Tools are an important component of any field of study. The following chapter elucidates the various tools that are related to material characterization.

Chapter 3- Microscopes are instruments designed to see elements that are too small to be seen with the naked eye. The most common types of microscopes are the optical microscope and the electron microscope. While optical microscope uses light as a means of magnification, electron microscopes takes the help of a beam of accelerated particles to achieve illumination. Electron microscopes can gain the resolution of 50 pm whereas optical microscopes can reach just 200 nm resolution. The chapter serves as a source to understand the major categories related to microscopes.

Chapter 4- Different types of microscopes use different techniques in analysis. One microscopy technique, transmission electron microscope, uses electrons to achieve higher level of resolution than a light microscope. It can see objects which are sized in angstroms. The chapter also explores the mechanism of scanning electron microscope. It is a type of electron microscope that uses focused beams to produce an image. Science and technology have undergone rapid developments in the past decade which has resulted in the discovery of significant techniques in microscopy; which have been extensively detailed in this chapter.

Indeed, my job was extremely crucial and challenging as I had to ensure that every chapter is informative and structured in a student-friendly manner. I am thankful for the support provided by my family and colleagues during the completion of this book.

Editor

1

Understanding Materials Characterization

The study, measure and analysis of a material's structure and properties is called as material characterization. Microscopic techniques such as the electron microscope and secondary ion mass spectrometry as well as macroscopic techniques like thermal analysis and density calculation are part of this process. This is an introductory chapter which will introduce briefly all the significant aspects of material characterization.

Materials

- Everything we use is made of some materials or a combination of materials. For examples, metals, ceramics, polymers, semiconductors and composites. We use these types of materials in our daily lives.
- It is well known that the properties of all these materials are determined by the materials' structure, which forms at various length scales as shown in figure.

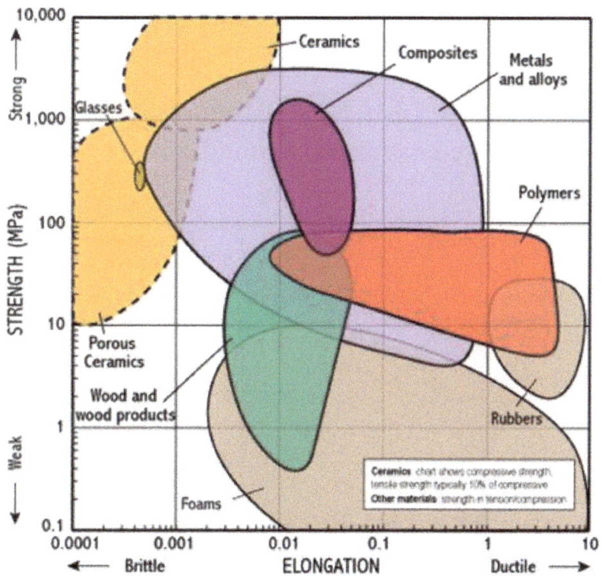

- Therefore, to understand the materials' structure at all possible relevant length scales, we apply all the basic principles of physics, chemistry and biology and see how this structure plays a key role in deciding the resulting properties.
- As the structural properties change with the fabrication process, the optimization of fabrication process with respect to different environments and characterization of materials forms crucial and vital aspects of materials research as shown in figure.

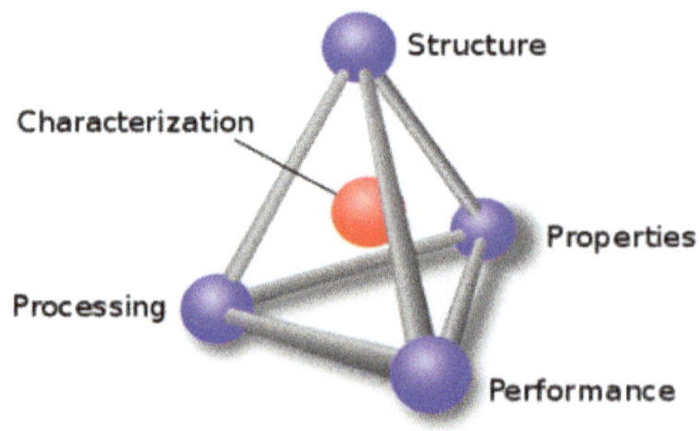

Schematic of various properties of materials.

- Furthermore, to design the scientific process of fabricating materials (as demonstrated in figure) and to meet the needs of the modern technology, understanding of various mutually interacting processes, mechanisms and materials using sophisticated instruments with an interdisciplinary approach is very much essential.

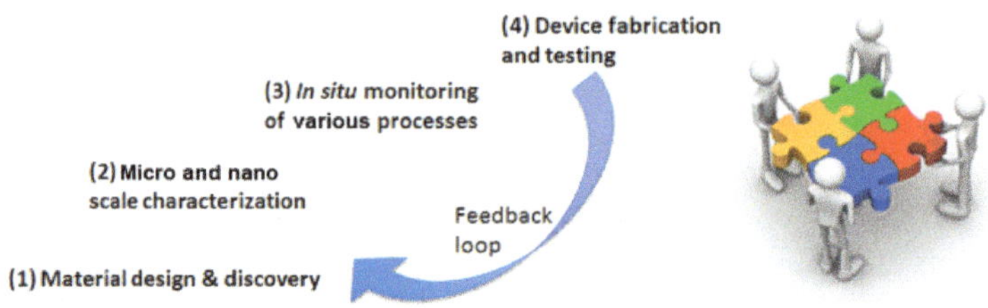

Schematic of the scientific process of fabricating materials.

- Therefore, the characterization of the materials has been one of the foundational pillars of materials science.
- The purpose of this course is to provide an introductory course on modern materials' characterizations, especially with respect to nanotechnology for both senior undergraduates and graduate students involved in interdisciplinary research.
- Although well established modern characterization techniques are available as depicted in figure, still many new techniques are evolving due to the emergency of nanotechnology field.

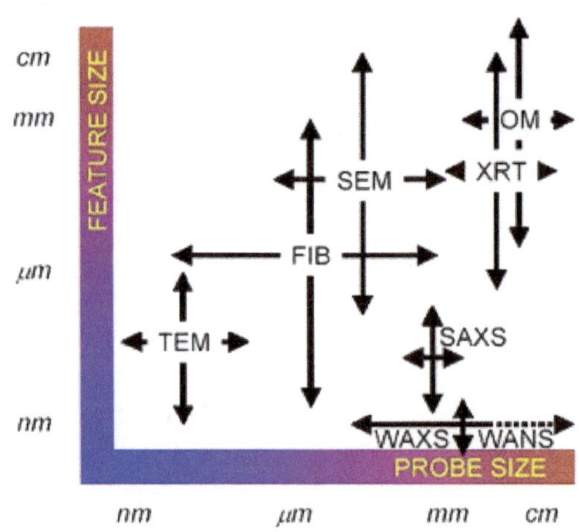

Schematic of feature size versus probe size and requirement of various instruments for structural characterization. OM - Optical Microscope, SEM - Scanning Electron Microscope, XRT - X-ray Tomography, FIB - Focused Ion Beam, TEM - Transmission Electron Microscope, SAXS - Small Angle X-ray Scattering, WAXS - Wide Angle X-ray Scattering, WANS - Wide Angle Neutron Scattering.

Characterization (Materials Science)

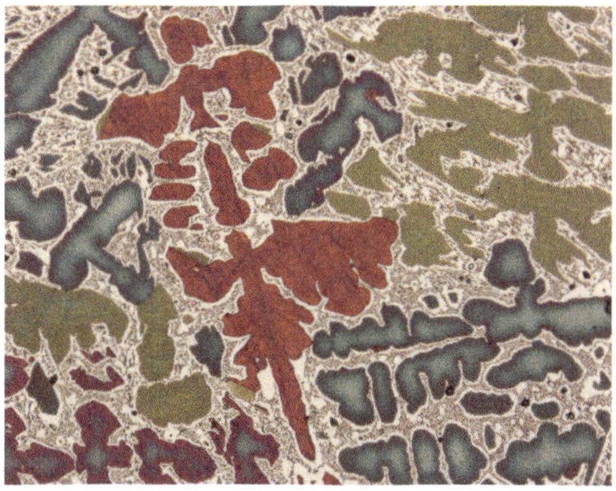

The characterization technique optical microscopy showing the micron scale dendritic microstructure of a bronze alloy.

Characterization, when used in materials science, refers to the broad and general process by which a material's structure and properties are probed and measured. It is a fundamental process in the field of materials science, without which no scientific understanding of engineering materials could be ascertained. The scope of the term often differs; some definitions limit the term's use to techniques which study the microscopic structure and properties of materials, while others use the term to refer to any materials analysis process including macroscopic techniques such as mechanical testing, thermal analysis and density calculation. The scale of the structures observed in materials characterization ranges from angstroms, such as in the imaging of individual atoms and chemical bonds, up to centimeters, such as in the imaging of coarse grain structures in metals.

While many characterization techniques have been practiced for centuries, such as basic optical microscopy, new techniques and methodologies are constantly emerging. In particular the advent of the electron microscope and Secondary ion mass spectrometry in the 20th century has revolutionized the field, allowing the imaging and analysis of structures and compositions on much smaller scales than was previously possible, leading to a huge increase in the level of understanding as to why different materials show different properties and behaviors. More recently, atomic force microscopy has further increased the maximum possible resolution for analysis of certain samples in the last 30 years.

Microscopy

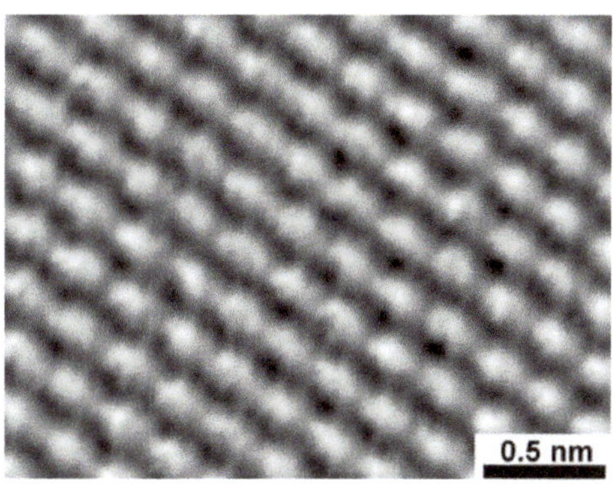

Image of a graphite surface at an atomic level obtained by an STM.

First X-ray diffraction view of Martian soil - CheMin analysis reveals feldspar, pyroxenes, olivine and more (Curiosity rover at "Rocknest", October 17, 2012).

Microscopy is a category of characterization techniques which probe and map the surface and sub-surface structure of a material. These techniques can use photons , electrons , ions or physical cantilever probes to gather data about a sample's structure on a range of length scales. Some common examples of microscopy instruments include:

- Optical Microscope
- Scanning Electron Microscope (SEM)
- Transmission Electron Microscope (TEM)
- Field Ion Microscope (FIM)
- Scanning Tunneling Microscope (STM)
- Scanning probe microscopy (SPM)
- Atomic Force Microscope (AFM)
- X-ray diffraction topography (XRT)

Spectroscopy

This group of techniques use a range of principles to reveal the chemical composition, composition variation, crystal structure and photoelectric properties of materials. Some common instruments include:

- Energy-dispersive X-ray spectroscopy (EDX)
- Wavelength dispersive X-ray spectroscopy (WDX)
- X-ray diffraction (XRD)
- Mass spectrometry
- Nuclear magnetic resonance spectroscopy (NMR)

- Secondary ion mass spectrometry (SIMS)
- Electron energy loss spectroscopy (EELS)
- Auger electron spectroscopy
- X-ray photoelectron spectroscopy (XPS)
- Ultraviolet-visible spectroscopy (UV-vis)
- Fourier transform infrared spectroscopy (FTIR)
- Thermoluminescence (TL)
- Photoluminescence (PL)
- Photon correlation spectroscopy/Dynamic light scattering (DLS)
- Terahertz spectroscopy
- Small-angle X-ray scattering (SAXS)
- Small-angle neutron scattering (SANS)
- X-ray Photon Correlation Spectroscopy (XPCS)

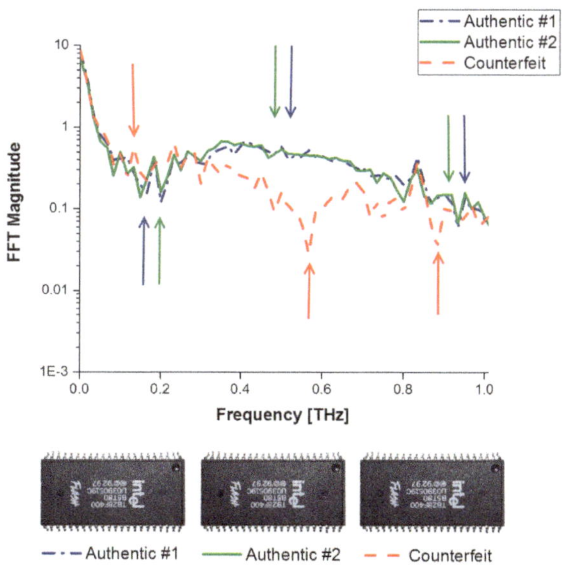

Material characterization using terahertz spectroscopy: the counterfeit IC is detected by revealing the differences between the constituent materials.

Macroscopic Testing

A huge range of techniques are used to characterize various macroscopic properties of materials, including:

- Mechanical testing, including tensile, compressive, torsional, creep, fatigue, toughness and hardness testing

- Differential thermal analysis (DTA)
- Dielectric thermal analysis
- Thermogravimetric analysis (TGA)
- Differential scanning calorimetry (DSC)
- Impulse excitation technique (IET)
- Ultrasound techniques, including resonant ultrasound spectroscopy and time domain ultrasonic testing methods

2

Microscope and its Types

Microscopes are instruments designed to see elements that are too small to be seen with the naked eye. The most common types of microscopes are the optical microscope and the electron microscope. While optical microscope uses light as a means of magnification, electron microscopes takes the help of a beam of accelerated particles to achieve illumination. Electron microscopes can gain the resolution of 50 pm whereas optical microscopes can reach just 200 nm resolution. The chapter serves as a source to understand the major categories related to microscopes.

Microscopes

- “Micro” refers to “tiny” and “Scope” refers to ‘view or look at”
- Microscopes are instruments designed to produce magnified visual or photographic images of objects too small to be seen with the naked eye and so they can be studied.
- Figure (from AWL Inc.,) shows typical examples of images at various levels and the types of microscopes needed to study the object of the interest.

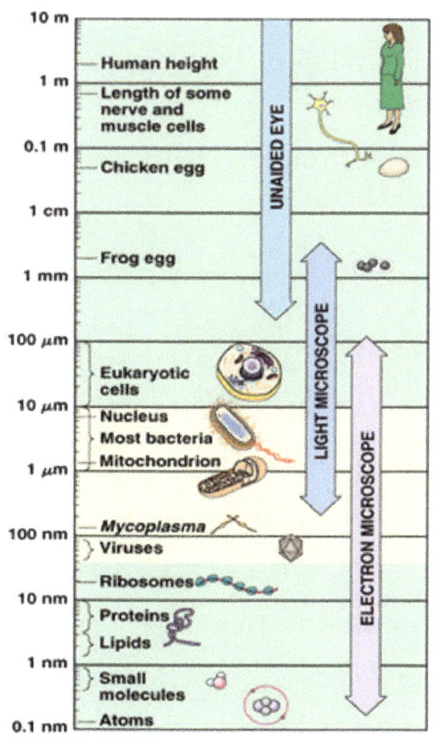

Typical examples of images at various levels and types of microscopes needed for analysis.

- While naked eye helps to view the objects down to 1 mm, light and electron microscopes are indeed needed to study the objects in the range 1 mm - 300 nm and 300 nm - 0.2 nm, respectively.
- Invention of the microscope is not credited to anyone. Actually, microscopes were byproducts of research on telescopes.
- First microscope: Van leeuwenhoek microscope (1600) shown in figure, very simple and able to study the structure of tissue, and was one of the principal scientists to discover protozoa in water.

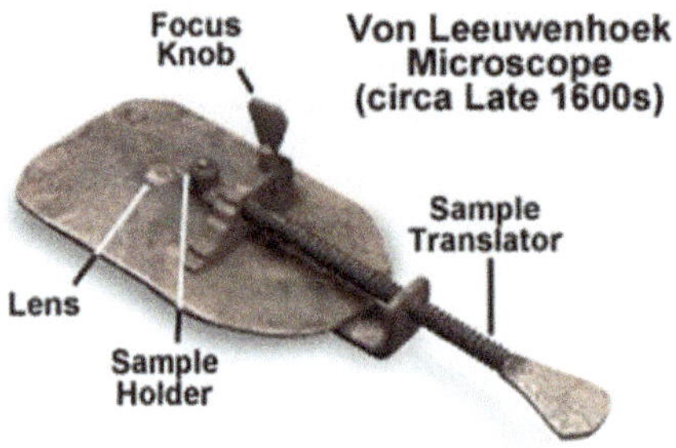

Typical view of Von Leeuwenhoek Microscope.

- Subsequently, the demand for microscopes for scientific use grew rapidly.
- Improvements in design came rapidly.
- Successful compound microscope by Zacharias Janessen (eye glass maker).
- Compound microscope was developed at the beginning of 1660's by Janssen brothers and Galileo.
- This early microscope suffered from chromatic (and spherical) aberration, and images were blurred and had colored halos.

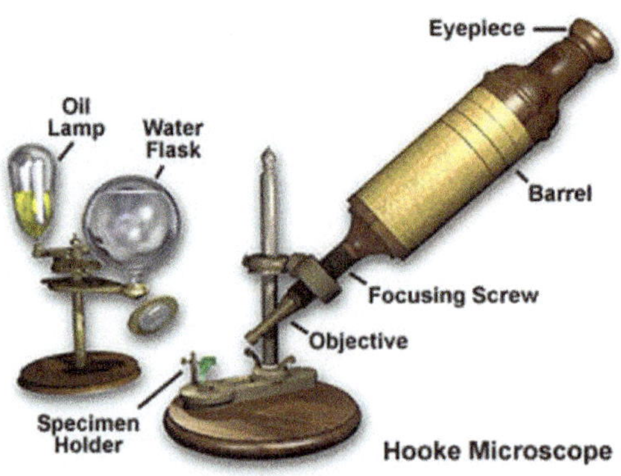

Photograph of Zacharias Janessen and Hooke microscope.

- Improvements: 1700 and 1800's - great improvement in microscopes
- Usually brass was the alloy of choice for the production of high-quality microscopes. But expensive.

- Tripod types structure was used to provide sturdy support for the microscope. This was considered one of the most advanced of its period.

Modern Microscopes: In 1900's, iron based structures were used to develope a microscope due to its cheaper cost; Only one eyepiece (monocular); Outside light source; reflected onto mirror; Very functional; Still used today.

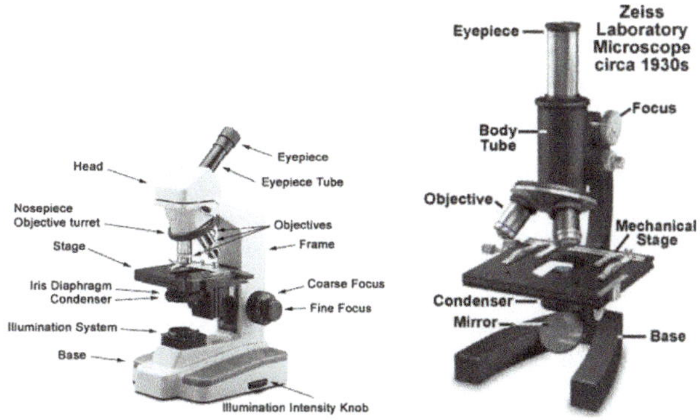

Photoview of single eye-piece microscopes.

Later on, Fancy type microscopes as shown in figure were developed (1998's). This provided

- Better images
- Large magnification
- Better lighting
- Easier to use
- Now everyone, at one time or another, has looked through a microscope.

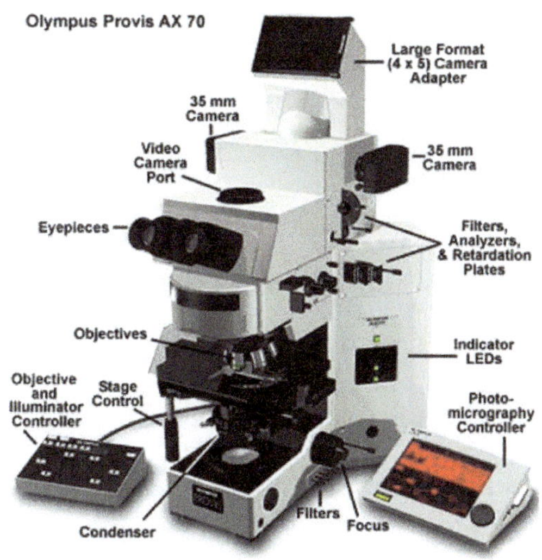

Photoview of single eye-piece microscopes.

- A microscope works very much like a telescope.
- A telescope must gather light from a dim, far away object as depicted in figure.

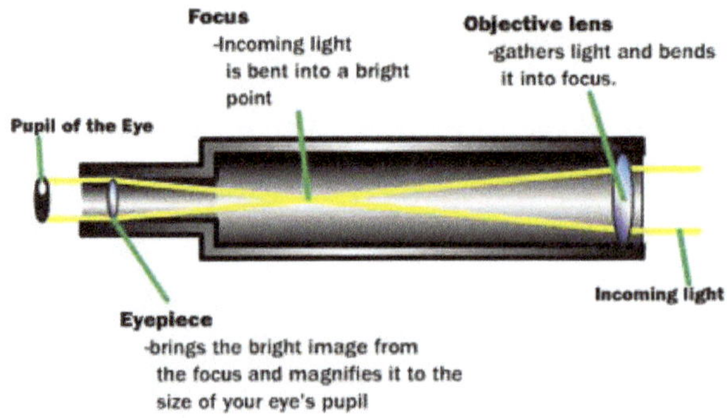

Schematic view of telescope working principle.

- So, it needs a large objective lens to gather as much light as possible and a long body to bring the image into focus.
- On the other hand, Unlike a telescope, a microscope must gather light from a tiny specimen that is located close-by as shown in figure.

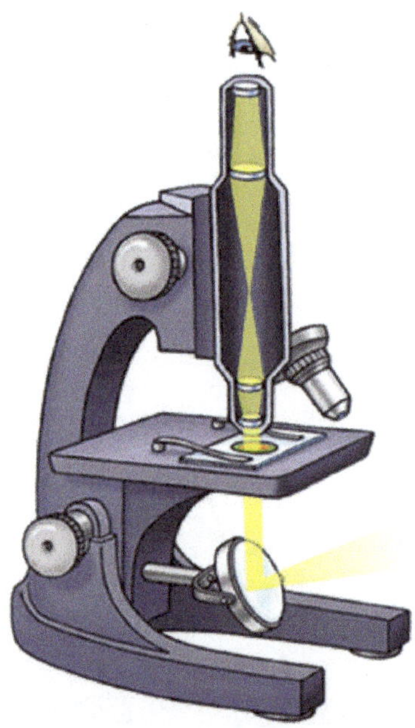

Schematic view of microscope.

- So the microscope does NOT need a large objective lens.
- Instead, the objective lens of a microscope is small.
- Then the image is again magnified by a second lens, called an eyepiece, as it is brought to your eye.

Need of Microscopes

- Several levels of structural information are of interest in research, process development, and reliability and failure analysis activities.

- The investigation of surface topography and morphology, including grain size and shape, grain boundaries, distribution of elements in the materials, existence of compounds, micro-cracking, formation of textured surfaces, etc., is more important to optimize the fabrication process.
- Furthermore, from industrial point of view, particularly from microelectronic device fabrication and optical coating technology, the cross-sectional views of multilayer structures where interfacial nature, substrate interactions, and geometry and perfection of electronic devices with associated conducting and insulating layers are very much essential.
- Therefore, microscopes are very much essential to understand the information down to atomic level and to optimize the fabrication process, resulting properties for commercializing the materials for products.

Concept of Resolution

As the resolution varies with the wavelength of the source utilized for the analysis, one needs to understand the concept of resolution.

Basic and Principles of Various Scatterings:

The interaction between source and matter must be understood clearly to obtain the essential information from the matter.

Understanding of the Instruments:

We shall briefly cover the major components of the different types of microscopes and the working principles of these components to understand the interaction processes which are taking places during the observation of the structural information.

- The image of an object can be magnified when viewed through a simple lens. By combining a number of lenses in a suitable manner, a microscope can be produced. This would yield very high magnification values.
- Resolving the smallest distance between two points, basically, referred as resolution or resolving power (Eye resolution: 0.1 ~ 0.2 mm).
- The instrument revealing details finer than 0.1 mm → Microscope.
- The Rayleigh criterion defines the resolution of a light microscope as:
- $\delta = 0.61 * [\lambda / (\mu . \sin\beta)]$

where λ is the wavelength of the radiation, μ is the refractive index of the view medium and β is the semi-angle of collection of the magnifying lens.

Taking green light as an example (λ = 550 nm), the typical resolution value turns out to be → 300 nm.

Optical Microscope

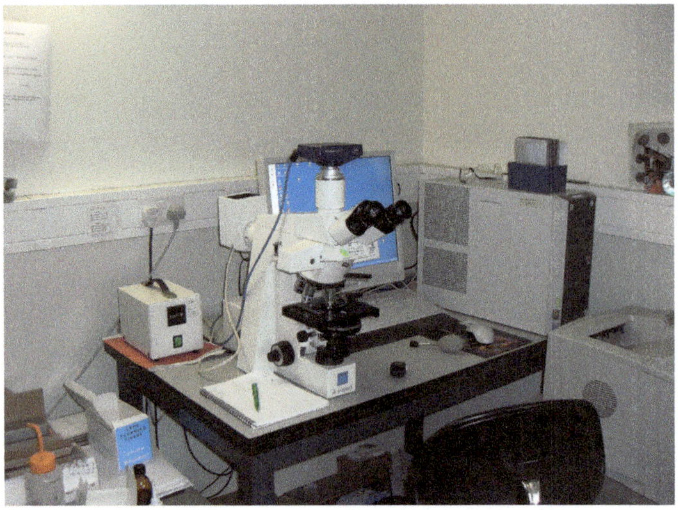

A modern microscope with a mercury bulb for fluorescence microscopy. The microscope has a digital camera, and is attached to a computer.

The optical microscope, often referred to as light microscope, is a type of microscope which uses visible light and a system of lenses to magnify images of small samples. Optical microscopes are the oldest design of microscope and were possibly invented in their present compound form in the 17th century. Basic optical microscopes can be very simple, although there are many complex designs which aim to improve resolution and sample contrast.

The image from an optical microscope can be captured by normal light-sensitive cameras to generate a micrograph. Originally images were captured by photographic film but modern developments in CMOS and charge-coupled device (CCD) cameras allow the capture of digital images. Purely digital microscopes are now available which use a CCD camera to examine a sample, showing the resulting image directly on a computer screen without the need for eyepieces.

Alternatives to optical microscopy which do not use visible light include scanning electron microscopy and transmission electron microscopy.

On 8 October 2014, the Nobel Prize in Chemistry was awarded to Eric Betzig, William Moerner and Stefan Hell for "the development of super-resolved fluorescence microscopy," which brings "optical microscopy into the nanodimension".

Types

There are two basic types of optical microscopes: simple microscopes and compound microscopes. A simple microscope is one which uses a single lens for magnification, such as a magnifying glass. A compound microscope uses several lenses to enhance the magnification of an object. The vast majority of modern research microscopes are compound microscopes while some cheaper commercial digital microscopes are simple single lens microscopes. Compound microscopes can be further divided into a variety of other types of microscopes which differ in their optical configurations, cost, and intended purposes.

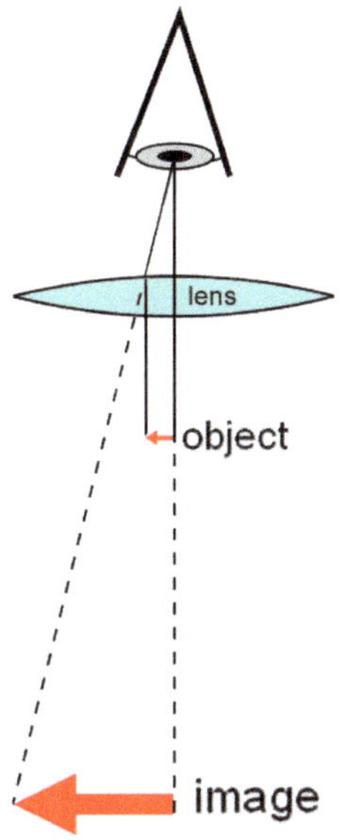

Diagram of a simple microscope

Simple Microscope

A simple microscope uses a lens or set of lenses to enlarge an object through angular magnification alone, giving the viewer an erect enlarged virtual image. The use of a single convex lens or groups of lenses are still found in simple magnification devices such as the magnifying glass, loupes, and eyepieces for telescopes and microscopes.

Compound Microscope

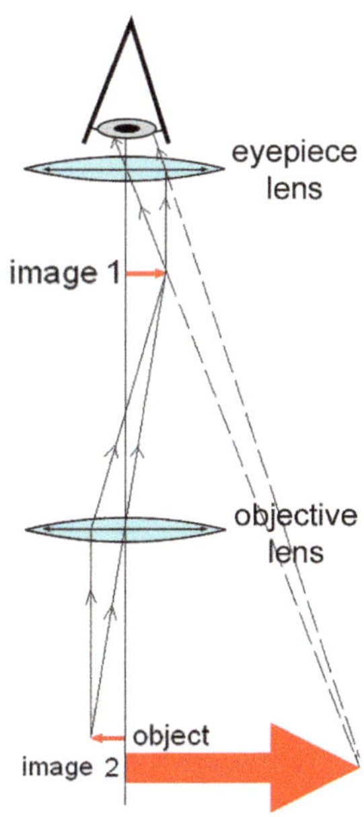

Diagram of a compound microscope

A compound microscope uses a lens close to the object being viewed to collect light (called the objective lens) which focuses a real image of the object inside the microscope (image 1). That image is then magnified by a second lens or group of lenses (called the eyepiece) that gives the viewer an enlarged inverted virtual image of the object (image 2). The use of a compound objective/eyepiece combination allows for much higher magnification. Common compound microscopes often feature exchangeable objective lenses, allowing the user to quickly adjust the magnification. A compound microscope also enables more advanced illumination setups, such as phase contrast.

Other Microscope Variants

There are many variants of the compound optical microscope design for specialized purposes. Some of these are physical design differences allowing specialization for certain purposes:

- Stereo microscope, a low-powered microscope which provides a stereoscopic view of the sample, commonly used for dissection.
- Comparison microscope, which has two separate light paths allowing direct comparison of two samples via one image in each eye.
- Inverted microscope, for studying samples from below; useful for cell cultures in liquid, or for metallography.
- Student microscope, designed for low cost, durability, and ease of use.
- Fiber optic connector inspection microscope, designed for connector end-face inspection

Other microscope variants are designed for different illumination techniques:

- Petrographic microscope, whose design usually includes a polarizing filter, rotating stage and gypsum plate to facilitate the study of minerals or other crystalline materials whose optical properties can vary with orientation.
- Polarizing microscope, similar to the petrographic microscope.
- Phase contrast microscope, which applies the phase contrast illumination method.
- Epifluorescence microscope, designed for analysis of samples which include fluorophores.
- Confocal microscope, a widely used variant of epifluorescent illumination which uses a scanning laser to illuminate a sample for fluorescence.
- Student microscope – an often low-power microscope with simplified controls and sometimes low quality optics designed for school use or as a starter instrument for children.

Digital Microscope

A digital microscope is a microscope equipped with a digital camera allowing observation of a sample via a computer. Microscopes can also be partly or wholly computer-controlled with various levels of automation. Digital microscopy allows greater analysis of a microscope image, for example measurements of distances and areas and quantitaton of a fluorescent or histological stain.

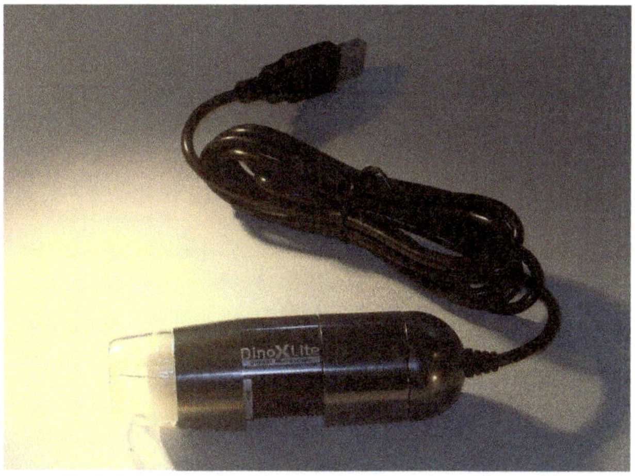

A miniature USB microscope.

Low-powered digital microscopes, USB microscopes, are also commercially available. These are essentially webcams with a high-powered macro lens and generally do not use transillumination. The camera attached directly to the USB port of a computer, so that the images are shown directly on the monitor. They offer modest magnifications (up to about 200×) without the need to use eyepieces, and at very low cost. High power illumination is usually provided by an LED source or sources adjacent to the camera lens.

Digital microscopy with very low light levels to avoid damage to vulnerable biological samples is available using sensitive photon-counting digital cameras. It has been demonstrated that a light source providing pairs of entangled photons may minimize the risk of damage to the most light-sensitive samples. In this application of ghost imaging to photon-sparse microscopy, the sample is illuminated with infrared photons, each of which is spatially correlated with an entangled partner in the visible band for efficient imaging by a photon-counting camera.

History

Invention

The earliest microscopes were single lens magnifying glasses with limited magnification which date at least as far back as the wide spread use of lenses in eyeglasses in the 13th century.

Compound microscopes first appeared in Europe around 1620 including one demonstrated by Cornelis Drebbel in London (around 1621) and one exhibited in Rome in 1624.

The actual inventor of the compound microscope is unknown although many claims have been made over the years. These include a claim 35 years after they appeared by Dutch spectacle-maker Johannes Zachariassen that his father, Zacharias Janssen, invented the compound microscope and/or the telescope as early as 1590. Johannes' (some claim dubious) testimony pushes the invention date so far back that Zacharias would have been a child at the time, leading to speculation that, for Johannes' claim to be true, the compound microscope would have to have been invented by Johannes' grandfather, Hans Martens. Another claim is that Janssen's competitor, Hans Lippershey (who applied for the first telescope patent in 1608) also invented the compound microscope. Other historians point to the Dutch innovator Cornelis Drebbel with his 1621 compound microscope.

Galileo Galilei is also sometimes cited as a compound microscope inventor. After 1610 he found that he could close focus his telescope to view small objects, such as flies, close up and/or could look through the wrong end in reverse to magnify small objects. After seeing the compound microscope exhibited in Rome in 1624, Galileo built his own improved version. In 1625 Giovanni Faber coined the name *microscope* for the compound microscope Galileo submitted to the Accademia dei Lincei in 1624 (Galileo had called it the "*occhiolino*" or "*little eye*"). Faber coined the name from the Greek words micron meaning "small", and skopein meaning "to look at", a name meant to be analogous with "telescope", another word coined by the Linceans.

Christiaan Huygens, another Dutchman, developed a simple 2-lens ocular system in the late 17th century that was achromatically corrected, and therefore a huge step forward in microscope development. The Huygens ocular is still being produced to this day, but suffers from a small field size, and other minor disadvantages.

Popularization

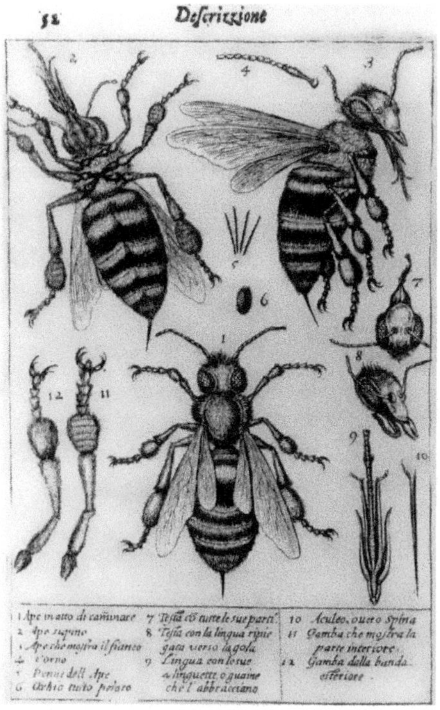

The oldest published image known to have been made with a microscope:
bees by Francesco Stelluti, 1630

Antonie van Leeuwenhoek (1632–1724) is credited with bringing the microscope to the attention of biologists, even though simple magnifying lenses were already being produced in the 16th century. Van Leeuwenhoek's home-made microscopes were simple microscopes, with a single very small, yet strong lens. They were awkward in use, but enabled van Leeuwenhoek to see detailed images. It took about 150 years of optical development before the compound microscope was able to provide the same quality image as van Leeuwenhoek's simple microscopes, due to difficulties in configuring multiple lenses. In the 1850s John Leonard Riddell, Professor of Chemistry at Tulane University, invented the first practical binocular microscope while carrying out one of the earliest and most extensive American microscopic investigations of cholera.

Lighting Techniques

While basic microscope technology and optics have been available for over 400 years it is much more recently that techniques in sample illumination were developed to generate the high quality images seen today.

In August 1893 August Köhler developed Köhler illumination. This method of sample illumination gives rise to extremely even lighting and overcomes many limitations of older techniques of sample illumination. Before development of Köhler illumination the image of the light source, for example a lightbulb filament, was always visible in the image of the sample.

The Nobel Prize in physics was awarded to Dutch physicist Frits Zernike in 1953 for his development of phase contrast illumination which allows imaging of transparent samples. By using interference rather than absorption of light, extremely transparent samples, such as live mammalian cells, can be imaged without having to use staining techniques. Just two years later, in 1955, Georges Nomarski published the theory for differential interference contrast microscopy, another interference-based imaging technique.

Fluorescence Microscopy

Modern biological microscopy depends heavily on the development of fluorescent probes for specific structures within a cell. In contrast to normal transilluminated light microscopy, in fluorescence microscopy the sample is illuminated through the objective lens with a narrow set of wavelengths of light. This light interacts with fluorophores in the sample which then emit light of a longer wavelength. It is this emitted light which makes up the image.

Since the mid 20th century chemical fluorescent stains, such as DAPI which binds to DNA, have been used to label specific structures within the cell. More recent developments include immunofluorescence, which uses fluorescently labelled antibodies to recognise specific proteins within a sample, and fluorescent proteins like GFP which a live cell can express making it fluorescent.

Components

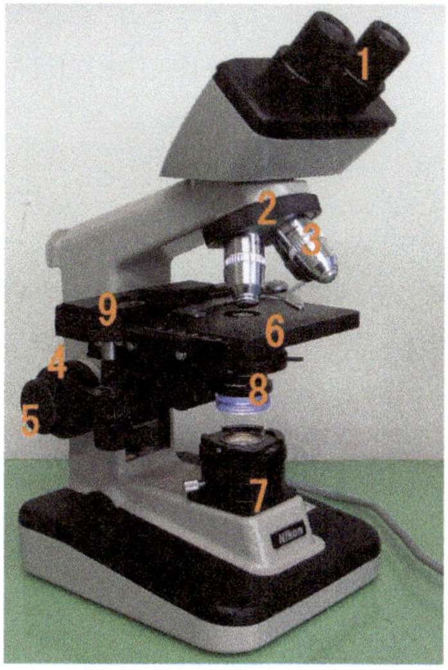

Basic optical transmission microscope elements (1990s)

All modern optical microscopes designed for viewing samples by transmitted light share the same basic components of the light path. In addition, the vast majority of microscopes have the same 'structural' components (numbered below according to the image on the right):

- Eyepiece (ocular lens) (1)
- Objective turret, revolver, or revolving nose piece (to hold multiple objective lenses) (2)
- Objective lenses (3)
- Focus knobs (to move the stage)
    - Coarse adjustment (4)
    - Fine adjustment (5)
- Stage (to hold the specimen) (6)
- Light source (a light or a mirror) (7)
- Diaphragm and condenser (8)
- Mechanical stage (9)

Eyepiece (Ocular Lens)

The eyepiece, or ocular lens, is a cylinder containing two or more lenses; its function is to bring the image into focus for the eye. The eyepiece is inserted into the top end of the body tube. Eyepieces are interchangeable and many different eyepieces can be inserted with different degrees of magnification. Typical magnification values for eyepieces include 5×, 10× (the most common), 15X and 20×. In some high performance microscopes, the optical configuration of the objective lens and eyepiece are matched to give the best possible optical performance. This occurs most commonly with apochromatic objectives.

Objective Turret (Revolver or Revolving Nose Piece)

Objective turret, revolver, or revolving nose piece is the part that holds the set of objective lenses. It allows the user to switch between objective lenses.

Objective

At the lower end of a typical compound optical microscope, there are one or more objective lenses that collect light from the sample. The objective is usually in a cylinder housing containing a glass single or multi-element compound lens. Typically there will be around three objective lenses screwed into a circular nose piece which may be rotated to select the required objective lens. These arrangements are designed to be parfocal, which means that when one changes from one lens to another on a microscope, the sample stays in focus. Microscope objectives are characterized by two parameters, namely, magnification and numerical aperture. The former typically ranges from 5× to 100× while the latter ranges from 0.14 to 0.7, corresponding to focal lengths of about 40 to 2 mm, respectively. Objective lenses with higher magnifications normally have a higher numerical aperture and a shorter depth of field in the resulting image. Some high performance objective lenses may require matched eyepieces to deliver the best optical performance.

Oil Immersion Objective

Two Leica oil immersion microscope objective lenses: 100× (left) and 40× (right)

Some microscopes make use of oil-immersion objectives or water-immersion objectives for greater resolution at high magnification. These are used with index-matching material such as immersion oil or water and a matched cover slip between the objective lens and the sample. The refractive index of the index-matching material is higher than air allowing the objective lens to have a larger numerical aperture (greater than 1) so that the light is transmitted from the specimen to the outer face of the objective lens with minimal refraction. Numerical apertures as high as 1.6 can be achieved. The larger numerical aperture allows collection of more light making detailed observation of smaller details possible. An oil immersion lens usually has a magnification of 40 to 100×.

Focus Knobs

Adjustment knobs move the stage up and down with separate adjustment for coarse and fine focusing. The same controls enable the microscope to adjust to specimens of different thickness. In older designs of microscopes, the focus adjustment wheels move the microscope tube up or down relative to the stand and had a fixed stage.

Frame

The whole of the optical assembly is traditionally attached to a rigid arm, which in turn is attached to a robust U-shaped foot to provide the necessary rigidity. The arm angle may be adjustable to allow the viewing angle to be adjusted.

The frame provides a mounting point for various microscope controls. Normally this will include controls for focusing, typically a large knurled wheel to adjust coarse focus, together with a smaller knurled wheel to control fine focus. Other features may be lamp controls and/or controls for adjusting the condenser.

Stage

The stage is a platform below the objective which supports the specimen being viewed. In the center of the stage is a hole through which light passes to illuminate the specimen. The stage usually has arms to hold slides (rectangular glass plates with typical dimensions of 25×75 mm, on which the specimen is mounted).

At magnifications higher than 100× moving a slide by hand is not practical. A mechanical stage, typical of medium and higher priced microscopes, allows tiny movements of the slide via control knobs that reposition the sample/slide as desired. If a microscope did not originally have a mechanical stage it may be possible to add one.

All stages move up and down for focus. With a mechanical stage slides move on two horizontal axes for positioning the specimen to examine specimen details.

Focusing starts at lower magnification in order to center the specimen by the user on the stage. Moving to a higher magnification requires the stage to be moved higher vertically for re-focus at the higher magnification and may also require slight horizontal specimen position adjustment. Horizontal specimen position adjustments are the reason for having a mechanical stage.

Due to the difficulty in preparing specimens and mounting them on slides, for children it's best to begin with prepared slides that are centered and focus easily regardless of the focus level used.

Light Source

Many sources of light can be used. At its simplest, daylight is directed via a mirror. Most microscopes, however, have their own adjustable and controllable light source – often a halogen lamp, although illumination using LEDs and lasers are becoming a more common provision.

Condenser

The condenser is a lens designed to focus light from the illumination source onto the sample. The condenser may also include other features, such as a diaphragm and/or filters, to manage the quality and intensity of the illumination. For illumination techniques like dark field, phase contrast and differential interference contrast microscopy additional optical components must be precisely aligned in the light path.

Magnification

The actual power or magnification of a compound optical microscope is the product of the powers of the ocular (eyepiece) and the objective lens. The maximum normal magnifications of the ocular and objective are 10× and 100× respectively, giving a final magnification of 1,000×.

Magnification and Micrographs

When using a camera to capture a micrograph the effective magnification of the image must take into account the size of the image. This is independent of whether it is on a print from a film negative or displayed digitally on a computer screen.

In the case of photographic film cameras the calculation is simple; the final magnification is the product of: the objective lens magnification, the camera optics magnification and the enlargement factor of the film print relative to the negative. A typical value of the enlargement factor is around 5× (for the case of 35mm film and a 15x10 cm (6×4 inch) print).

In the case of digital cameras the size of the pixels in the CMOS or CCD detector and the size of the pixels on the screen have to be known. The enlargement factor from the detector to the pixels on screen can then be calculated. As with a film camera the final magnification is the product of: the objective lens magnification, the camera optics magnification and the enlargement factor.

Operation

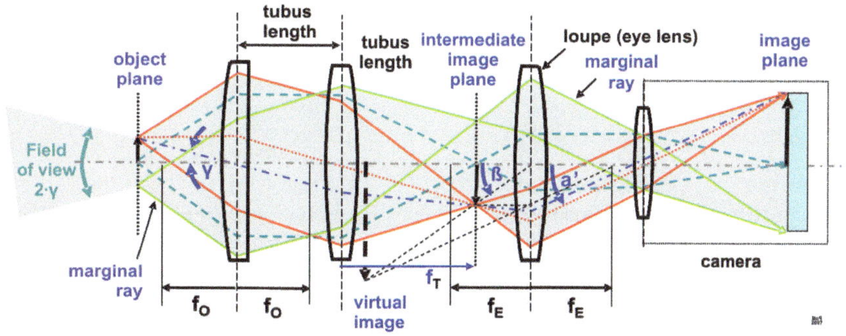

Optical path in a typical microscope

The optical components of a modern microscope are very complex and for a microscope to work well, the whole optical path has to be very accurately set up and controlled. Despite this, the basic operating principles of a microscope are quite simple.

The objective lens is, at its simplest, a very high-powered magnifying glass, i.e. a lens with a very short focal length. This is brought very close to the specimen being examined so that the light from the specimen comes to a focus about 160 mm inside the microscope tube. This creates an enlarged image of the subject. This image is inverted and can be seen by removing the eyepiece and placing a piece of tracing paper over the end of the tube. By carefully focusing a brightly lit specimen, a highly enlarged image can be seen. It is this real image that is viewed by the eyepiece lens that provides further enlargement.

In most microscopes, the eyepiece is a compound lens, with one component lens near the front and one near the back of the eyepiece tube. This forms an air-separated couplet. In many designs, the virtual image comes to a focus between the two lenses of the eyepiece, the first lens bringing the real image to a focus and the second lens enabling the eye to focus on the virtual image.

In all microscopes the image is intended to be viewed with the eyes focused at infinity (mind that the position of the eye in the above figure is determined by the eye's focus). Headaches and tired eyes after using a microscope are usually signs that the eye is being forced to focus at a close distance rather than at infinity.

Illumination Techniques

Many techniques are available which modify the light path to generate an improved contrast image from a sample. Major techniques for generating increased contrast from the sample include cross-polarized light, dark field, phase contrast and differential interference contrast illumination. A recent technique (Sarfus) combines cross-polarized light and specific contrast-enhanced slides for the visualization of nanometric samples.

Four examples of transilumination techniques used to generate contrast in a sample of tissue paper. 1.559 μm/pixel.

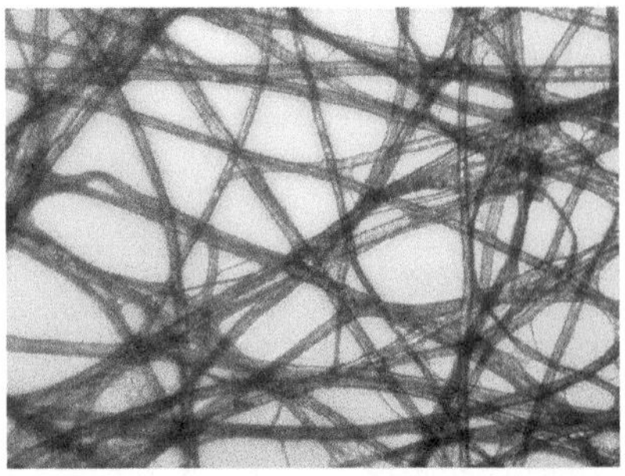

Bright field illumination, sample contrast comes from absorbance of light in the sample.

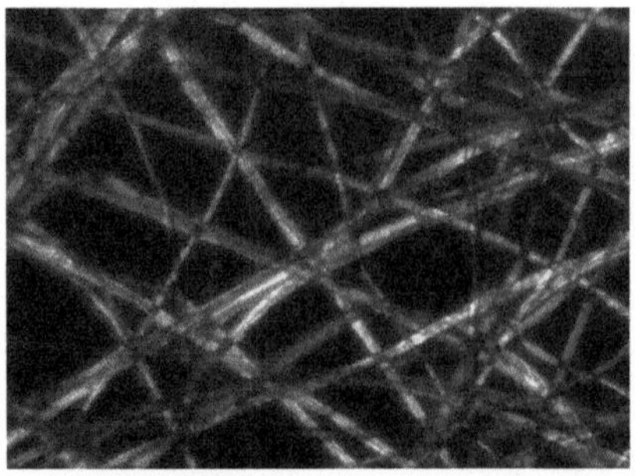

Cross-polarized light illumination, sample contrast comes from rotation of polarized light through the sample.

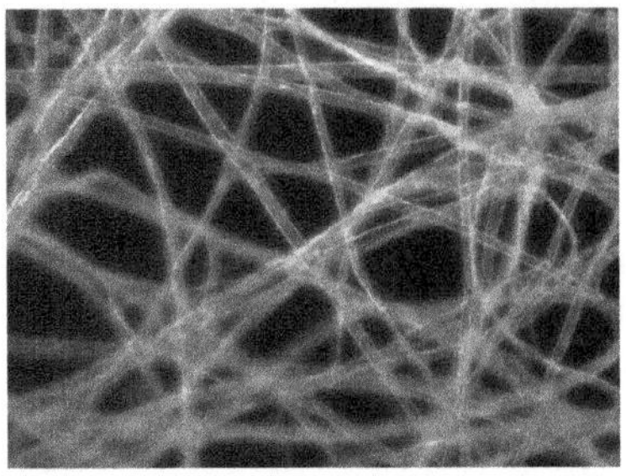

Dark field illumination, sample contrast comes from light scattered by the sample.

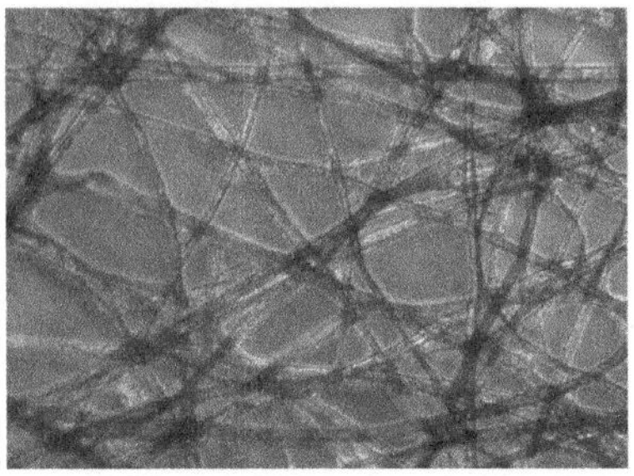

Phase contrast illumination, sample contrast comes from interference of different path lengths of light through the sample.

Other Techniques

Modern microscopes allow more than just observation of transmitted light image of a sample; there are many techniques which can be used to extract other kinds of data. Most of these require additional equipment in addition to a basic compound microscope.

- Reflected light, or incident, illumination (for analysis of surface structures)
- Fluorescence microscopy, both:
    1. Epifluorescence microscopy
    2. Confocal microscopy
- Microspectroscopy (where a UV-visible spectrophotometer is integrated with an optical microscope)
- Ultraviolet microscopy
- Near-Infrared microscopy
- Multiple transmission microscopy for contrast enhancement and aberration reduction.
- Automation (for automatic scanning of a large sample or image capture)

Applications

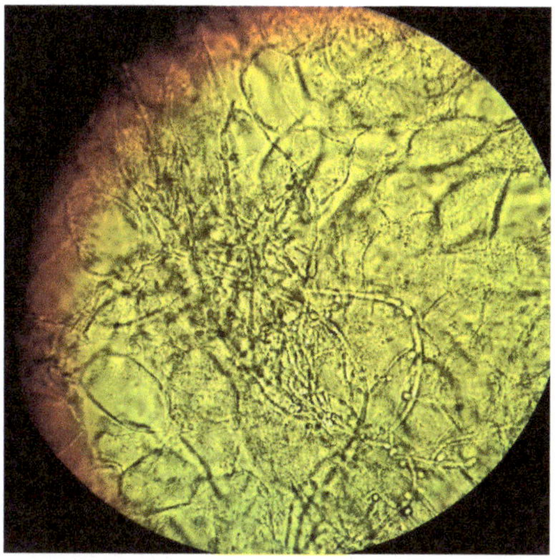

A 40x magnification image of cells in a medical smear test taken through an optical microscope using a wet mount technique, placing the specimen on a glass slide and mixing with a salt solution

Optical microscopy is used extensively in microelectronics, nanophysics, biotechnology, pharmaceutic research, mineralogy and microbiology.

Optical microscopy is used for medical diagnosis, the field being termed histopathology when dealing with tissues, or in smear tests on free cells or tissue fragments.

In industrial use, binocular microscopes are common. Aside from applications needing true depth perception, the use of dual eyepieces reduces eye strain associated with long workdays at a microscopy station. In certain applications, long-working-distance or long-focus microscopes are beneficial. An item may need to be examined behind a window, or industrial subjects may be a hazard to the objective. Such optics resemble telescopes with close-focus capabilities.

Measuring microscopes are used for precision measurement. There are two basic types. One has a reticle graduated to allow measuring distances in the focal plane. The other (and older) type has simple crosshairs and a micrometer mechanism for moving the subject relative to the microscope.

Limitations

At very high magnifications with transmitted light, point objects are seen as fuzzy discs surrounded by diffraction rings. These are called Airy disks. The *resolving power* of a microscope is taken as the ability to distinguish between two closely spaced Airy disks (or, in other words the ability of the microscope to reveal adjacent structural detail as distinct and separate). It is these impacts of diffraction that limit the ability to resolve fine details. The extent and magnitude of the diffraction patterns are affected by both the wavelength of light (λ), the refractive materials used to manufacture the objective lens and the numerical aperture (NA) of the objective lens. There is therefore a finite limit beyond which it is impossible to resolve separate points in the objective field, known as the diffraction limit. Assuming that optical aberrations in the whole optical set-up are negligible, the resolution *d*, can be stated as:

$$d = \frac{\lambda}{2NA}$$

Usually a wavelength of 550 nm is assumed, which corresponds to green light. With air as the external medium, the highest practical *NA* is 0.95, and with oil, up to 1.5. In practice the lowest value of *d* obtainable with conventional lenses is about 200 nm. A new type of lens using multiple scattering of light allowed to improve the resolution to below 100 nm.

Surpassing the Resolution Limit

Multiple techniques are available for reaching resolutions higher than the transmitted light limit described above. Holographic techniques, as described by Courjon and Bulabois in 1979, are also capable of breaking this resolution limit, although resolution was restricted in their experimental analysis.

Using fluorescent samples more techniques are available. Examples include Vertico SMI, near field scanning optical microscopy which uses evanescent waves, and stimulated emission depletion. In 2005, a microscope capable of detecting a single molecule was described as a teaching tool.

Despite significant progress in the last decade, techniques for surpassing the diffraction limit remain limited and specialized.

While most techniques focus on increases in lateral resolution there are also some techniques which aim to allow analysis of extremely thin samples. For example, sarfus methods place the thin sample on a contrast-enhancing surface and thereby allows to directly visualize films as thin as 0.3 nanometers.

Structured Illumination SMI

SMI (spatially modulated illumination microscopy) is a light optical process of the so-called point spread function (PSF) engineering. These are processes which modify the PSF of a microscope in a suitable manner to either increase the optical resolution, to maximize the precision of distance measurements of fluorescent objects that are small relative to the wavelength of the illuminating light, or to extract other structural parameters in the nanometer range.

Localization Microscopy SPDMphymod

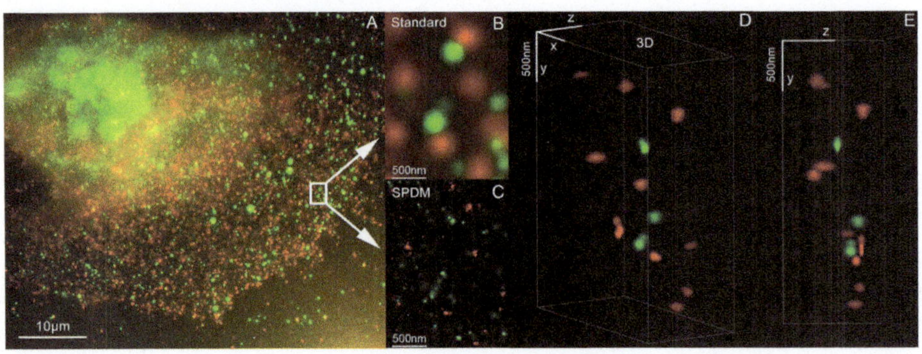

3D dual color super resolution microscopy with Her2 and Her3 in breast cells, standard dyes: Alexa 488, Alexa 568 LIMON

SPDM (spectral precision distance microscopy), the basic localization microscopy technology is a light optical process of fluorescence microscopy which allows position, distance and angle measurements on "optically isolated" particles (e.g. molecules) well below the theoretical limit of resolution for light microscopy. "Optically isolated" means that at a given point in time, only a single particle/molecule within a region of a size determined by conventional optical resolution (typically approx. 200–250 nm diameter) is being registered. This is possible when molecules within such a region all carry different spectral markers (e.g. different colors or other usable differences in the light emission of different particles).

Many standard fluorescent dyes like GFP, Alexa dyes, Atto dyes, Cy2/Cy3 and fluorescein molecules can be used for localization microscopy, provided certain photo-physical conditions are present. Using this so-called SPDMphymod (physically modifiable fluorophores) technology a single laser wavelength of suitable intensity is sufficient for nanoimaging.

3D Super Resolution Microscopy

3D super resolution microscopy with standard fluorescent dyes can be achieved by combination of localization microscopy for standard fluorescent dyes SPDMphymod and structured illumination SMI.

STED

Stimulated emission depletion (STED) microscopy image of actin filaments within a cell.

Stimulated emission depletion is a simple example of how higher resolution surpassing the diffraction limit is possible, but it has major limitations. STED is a fluorescence microscopy technique which uses a combination of light pulses to induce fluorescence in a small sub-population of fluorescent molecules in a sample. Each molecule produces a diffraction-limited spot of light in the image, and the centre of each of these spots corresponds to the location of the molecule. As the number of fluorescing molecules is low the spots of light are unlikely to overlap and therefore can be placed accurately. This process is then repeated many times to generate the image. Stefan Hell of the Max Planck Institute for Biophysical Chemistry was awarded the 10th German Future Prize in 2006 and Nobel Prize for Chemistry in 2014 for his development of the STED microscope and associated methodologies.

Alternatives

In order to overcome the limitations set by the diffraction limit of visible light other microscopes have been designed which use other waves.

- Atomic force microscope (AFM)
- Scanning electron microscope (SEM)
- Scanning ion-conductance microscopy (SICM)
- Scanning tunneling microscope (STM)
- Transmission electron microscopy (TEM)
- Ultraviolet microscope
- X-ray microscope

It is important to note that higher frequency waves have limited interaction with matter, for example soft tissues are relatively transparent to X-rays resulting in distinct sources of contrast and different target applications.

The use of electrons and X-rays in place of light allows much higher resolution – the wavelength of the radiation is shorter so the diffraction limit is lower. To make the short-wavelength probe non-destructive, the atomic beam imaging system (atomic nanoscope) has been proposed and widely discussed in the literature, but it is not yet competitive with conventional imaging systems.

STM and AFM are scanning probe techniques using a small probe which is scanned over the sample surface. Resolution in these cases is limited by the size of the probe; micromachining techniques can produce probes with tip radii of 5–10 nm.

Additionally, methods such as electron or X-ray microscopy use a vacuum or partial vacuum, which limits their use for live and biological samples (with the exception of an environmental scanning electron microscope). The specimen chambers needed for all such instruments also limits sample size, and sample manipulation is more difficult. Color cannot be seen in images made by these methods, so some information is lost. They are however, essential when investigating molecular or atomic effects, such as age hardening in aluminium alloys, or the microstructure of polymers.

Figure displays the electromagnetic spectrum over a wide range of wavelength and the extended region of a visible region lies in the range of 400 nm to 700 nm.

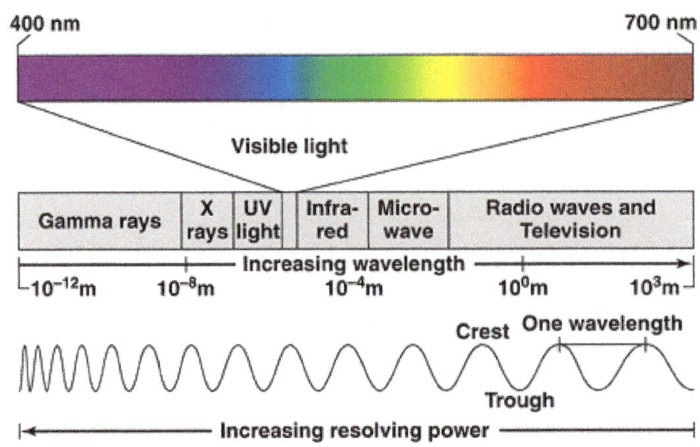

Typical electromagnetic spectrum highlighting the range of visible light.

Light Microscope: Definition

Optical or light microscopy as shown in the figure involves passing visible light transmitted through or reflected from the sample through a single or multiple lenses to allow a magnified view of the sample.

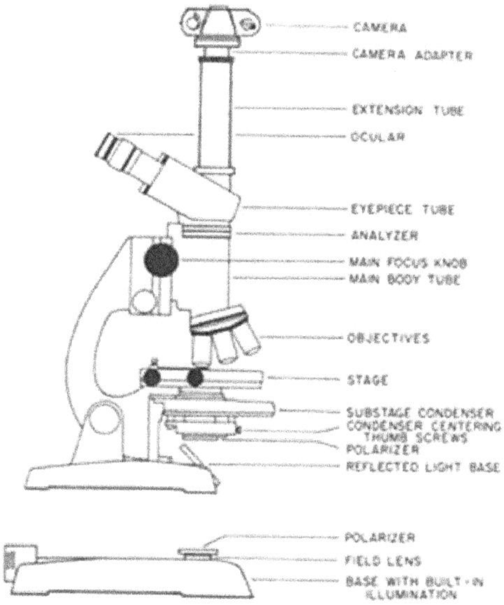

Schematic of light microscope.

The resulting image can be detected directly by the eye and/or imaged on a photographic plate or captured digitally.

The single lens with its attachments, or the system of lenses and imaging equipment, along with the appropriate lighting equipment, sample stage and support, makes up the basic light microscope.

Light Microscope: Magnification

Magnification of the Microscope

M of Microscope = M of Objective x M of Eyepiece x M of Intermediate Factor

- For Example:

Objective = 60 x

Eyepiece = 10 x

Intermediate Factor = 1 x

Overall M turns out to be 600 x

The magnification of the compound microscope is given as $M = m_o M_e$

Where, Me is angular magnification of eyepiece given as 25/fe and mo is linear magnification of objective defined as -(L / fo). fo and fe are the focal lengths of objective and eye-piece, respectively as given in figure.

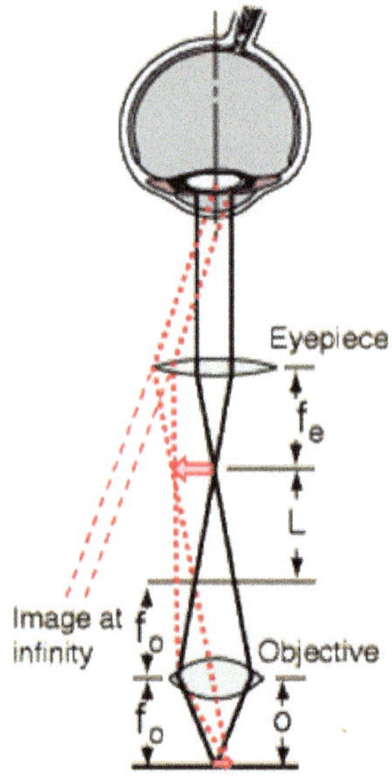

Schematic of viewing image using light microscope.

Light Microscope: Types

There are many types of light microscopes for imaging purposes:

- Bright-field microscope
- Dark-field microscope
- Phase-contrast microscope
- Fluorescence microscope

Light Microscope: Bright-field

Bright-field microscopy is the simplest microscope of all the types of optical microscopes. Sample is illuminated using white light from bottom side as shown in figure and the image is observed from above. The contrast in the specimen is caused by absorbance of some of the transmitted light in dense areas of the sample.

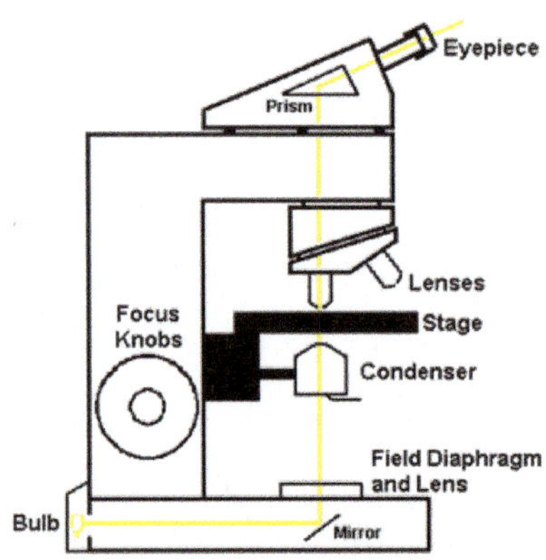

Schematic of light microscope.

The typical appearance of a bright-field microscopy image is a dark sample on a bright background and hence the name.

Series of lenses are placed for suitable magnification. It has one or few ocular lenses to view the sample image.

- Total magnification = Magnification of objective lens x Magnification of an ocular lens
- Most have the condenser lens (direct light through specimen)
- Oil immersion lens improves the light through objective lens and increases resolution, as shown in figure. Oil immersion is mainly used to increase the resolution of a microscope. This is obtained by immersing both the objective lens and the specimen in a transparent oil of high refractive index (~ 1.515) and specific optical and viscosity characteristics. This helps to increase the numerical aperture of the objective lens.

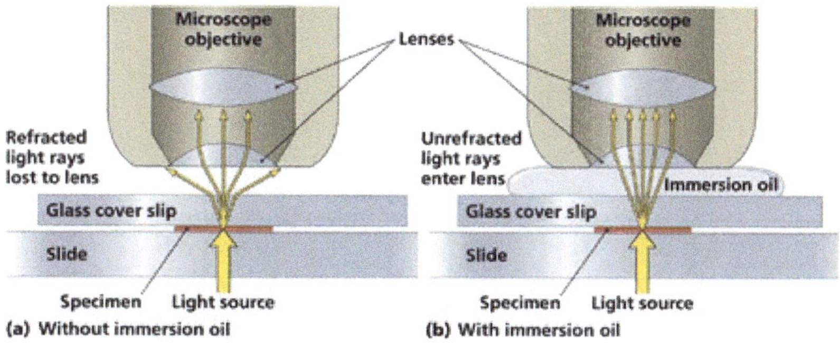

Light microscope with and without oil immersion lens.

Bright field is the most universal technique used in light microscope. Typical image observed for Hemlock leaf using the light microscope is shown in figure.

Optical image of Hemlock leaf.

Light Microscope: Dark field

Figure shows the light path in a dark-field microscope. This mode is best for observing the pale objects. Only light rays scattered by specimen enter objective lens and the specimen appears light against dark background.

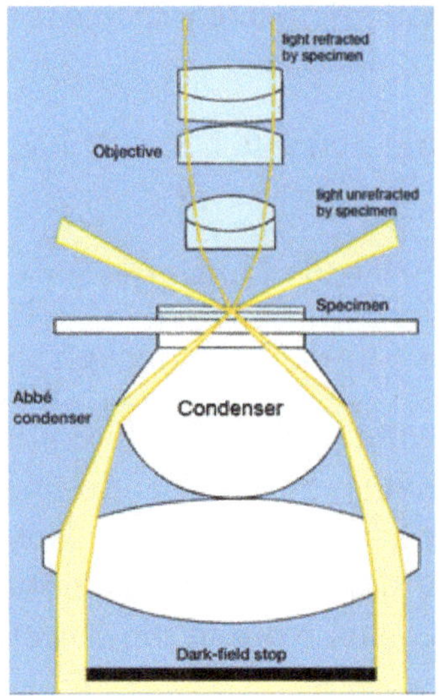

Schematic of light path for dark field imaging.

It produces a bright image of the object against a dark background. The contrast increases, which enables observation of more details as shown in the figure, which are not seen in the bright background. This mode is used to observe living, unstained preparations.

Microscope image under dark field imaging condition.

Fine structures can often not be seen in front of a bright background

Light Microscope: Phase contrast

Phase contrast microscopy is an optical microscopy technique that converts phase shifts in light passing through a transparent specimen to brightness changes in the image. Phase shifts are invisible, but become visible when shown as brightness variations.

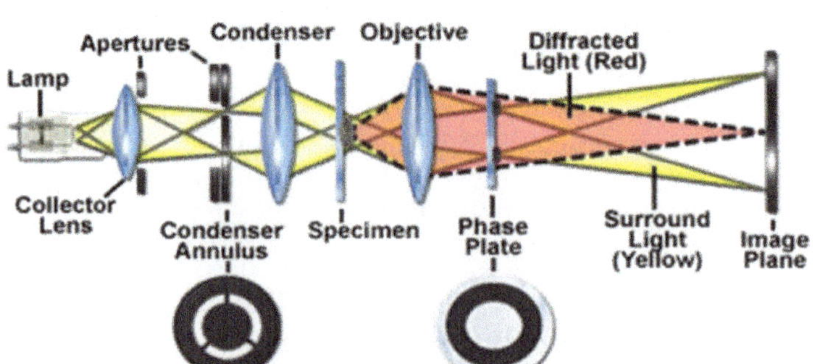

Schematic of Phase cntrast light microscope.

The basic principle to make phase changes visible in phase contrast microscopy is to separate the illuminating background light from the specimen scattered light. This makes up the foreground details, and to manipulate these differently.

Phase contrast microscopy is particularly important in biology. It reveals many cellular structures (right image) that are not visible with a simpler bright-field microscope (left image), as shown in the figure.

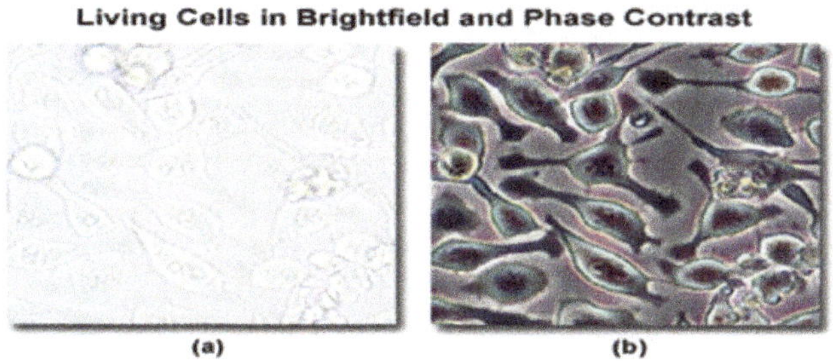

Images of Cellular structure obtained using (a) bright-field and (b) phase contrast microscopes.

- Light rays in phase produce brighter images, while light rays out of phase produce darker images.
- Contrast is created because light waves are out of phase.
- Enhances the contrast between intracellular structures having slight differences in the refractive index.
- Used to examine living organisms or specimens that would be damaged/altered by attaching them to slides or staining. Excellent way to observe living cells.

Two Types

- Phase-contrast microscope
- Differential interference contrast microscope

Light Microscope: Differential Interference Contrast Microscope

Differential interference contrast (DIC) microscopy also known as Nomarski microscopy (NM) is an optical microscopy illumination technique used to enhance the contrast in unstained and transparent samples. It works on the principle of interferometry to gain information about the optical path length of the sample. A relatively complex lighting scheme produces an image with the object appearing black to white on a grey background.

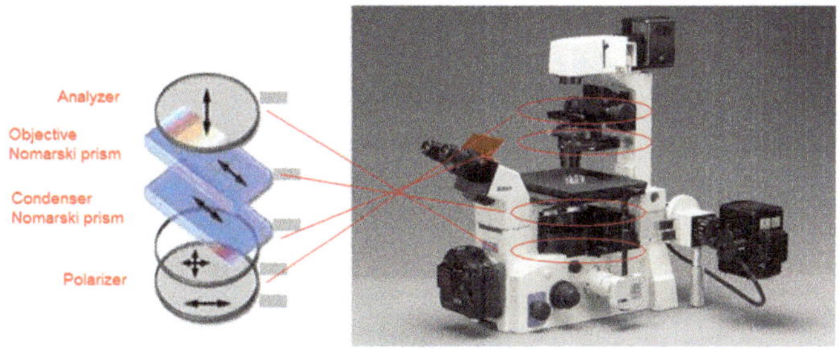

Schematic of DIC microscopy principle and set up.

This image is similar to that obtained by phase contrast microscopy but without the bright diffraction halo.

It creates an image by detecting differences in refractive indices and thickness of distinctive parts of specimen. This is an excellent way to observe living cells.

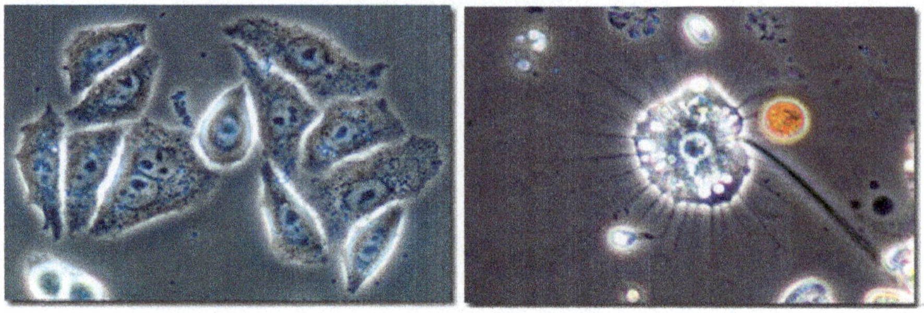

Images obtained using DIC microscope for HeLa Cell structure (left) and Heliozoans (right).

Light Microscope: The Fluorescence Microscope

The absorption and subsequent re-radiation of light by organic and inorganic specimens are typically the result of well-established physical phenomena described as being either fluorescence or phosphorescence.

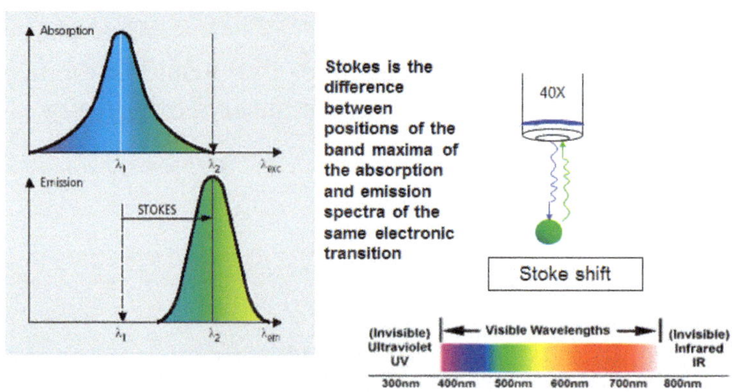

Typical absorption and emission spectra and principle of stoke shift.

- The fluorescence microscopy is an essential tool particularly in biology, the biomedical sciences and in materials science due to attributes that are not readily available in other contrast modes with traditional optical microscopy.
- In this case, the specimen is exposed to ultraviolet, violet and / or blue lights.
- The application of an array of fluorochromes has made it possible to identify cells and sub-microscopic cellular components with a high degree of specificity amid non-fluorescing material.
- This reveals bright images of the object resulting from the fluorescent light emitted by the specimen. In fact, this microscope is capable of revealing the presence of a single molecule.

- Although this cannot provide a spatial resolution below the diffraction limit of specific specimen features, the detection of fluorescing molecules below such limits is always achieved.

Light Microscope: The Fluorescence Microscope: Principle

Figure shows the typical diagram of fluorescent microscope and its typical components: light source (Xenon arc, mercury-vapor lamp and lasers), the excitation filter, the dichroic mirror and the emission filter.

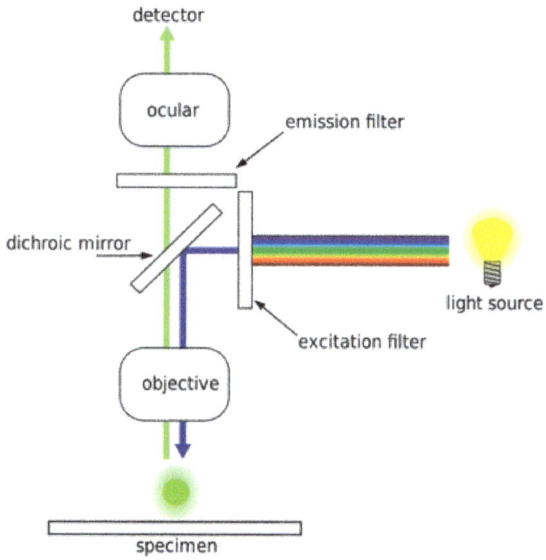

Schematic of fluorescence microscope.

- Specimen is illuminated with light of a specific wavelength.
- Light is absorbed by the fluorophores causing an emission of longer wavelengths.
- The illumination light is separated from weaker emitted fluorescence through the use of a spectral emission filter.

Light Microscope: The Fluorescence Microscope

Figure displays the typical images of cells obtained using the fluorescence microscope.

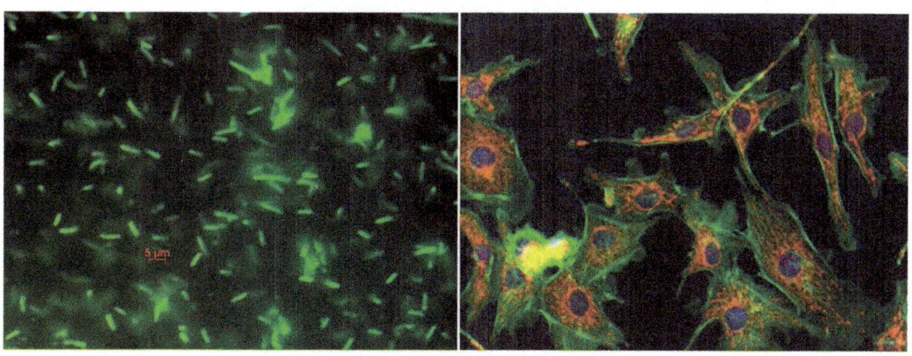

Fluorescence microscope image of RV308 cells electroporated with fluorescein labeled oligonucleotides (left) and cells (right).

Electron Microscope

An electron microscope is a microscope that uses a beam of accelerated electrons as a source of illumination. As the wavelength of an electron can be up to 100,000 times shorter than that of visible light photons, electron microscopes have a higher resolving power than light microscopes and can reveal the structure of smaller objects. A transmission electron microscope can achieve better than 50 pm resolution and magnifications of up to about 10,000,000x whereas most light microscopes are limited by diffraction to about 200 nm resolution and useful magnifications below 2000x.

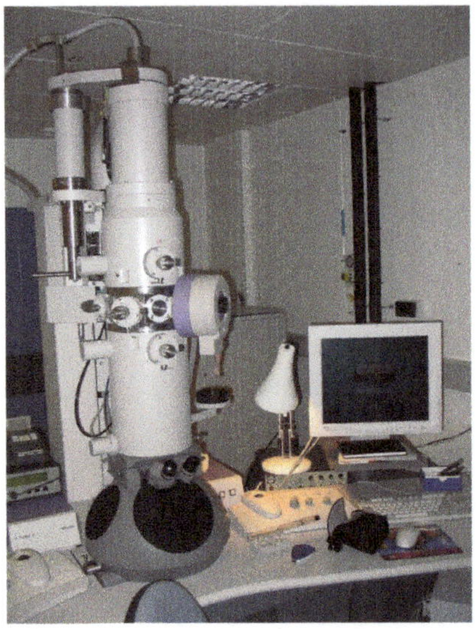

A modern transmission electron microscope

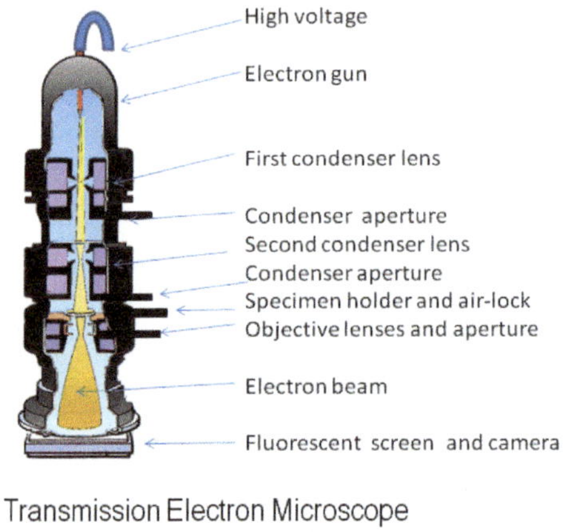

Diagram of a transmission electron microscope

Transmission electron microscopes use electrostatic and electromagnetic lenses to control the electron beam and focus it to form an image. These electron optical lenses are analogous to the glass lenses of an optical light microscope.

Electron microscope constructed by Ernst Ruska in 1933

Electron microscopes are used to investigate the ultrastructure of a wide range of biological and inorganic specimens including microorganisms, cells, large molecules, biopsy samples, metals, and crystals. Industrially, electron microscopes are often used for quality control and failure analysis. Modern electron microscopes produce electron micrographs using specialized digital cameras and frame grabbers to capture the image.

History

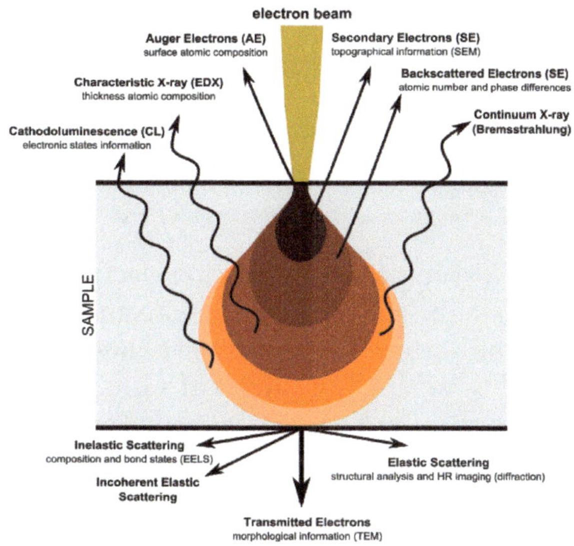

Diagram illustrating the phenomena resulting of the interaction of highly energetic electrons with matter.

The first electromagnetic lens was developed in 1926 by Hans Busch.

According to Dennis Gabor, the physicist Leó Szilárd tried in 1928 to convince Busch to build an electron microscope, for which he had filed a patent.

The physicist Ernst Ruska and the electrical engineer Max Knoll constructed the prototype electron microscope in 1931, capable of four-hundred-power magnification; the apparatus was the first demonstration of the principles of electron microscopy. Two years later, in 1933, Ruska built an electron microscope that exceeded the resolution attainable with an optical (light) microscope. Moreover, Reinhold Rudenberg, the scientific director of Siemens-Schuckertwerke, obtained the patent for the electron microscope in May 1931.

In 1932, Ernst Lubcke of Siemens & Halske built and obtained images from a prototype electron microscope, applying concepts described in the Rudenberg patent applications. Five years later (1937), the firm financed the work of Ernst Ruska and Bodo von Borries, and employed Helmut Ruska (Ernst's brother) to develop applications for the microscope, especially with biological specimens. Also in 1937, Manfred von Ardenne pioneered the scanning electron microscope. The first commercial electron microscope was produced in 1938 by Siemens. The first North American electron microscope was constructed in 1938, at the University of Toronto, by Eli Franklin Burton and students Cecil Hall, James Hillier, and Albert Prebus; and Siemens produced a transmission electron microscope (TEM) in 1939. Although contemporary transmission electron microscopes are capable of two million-power magnification, as scientific instruments, they remain based upon Ruska's prototype.

Types

Transmission Electron Microscope (Tem)

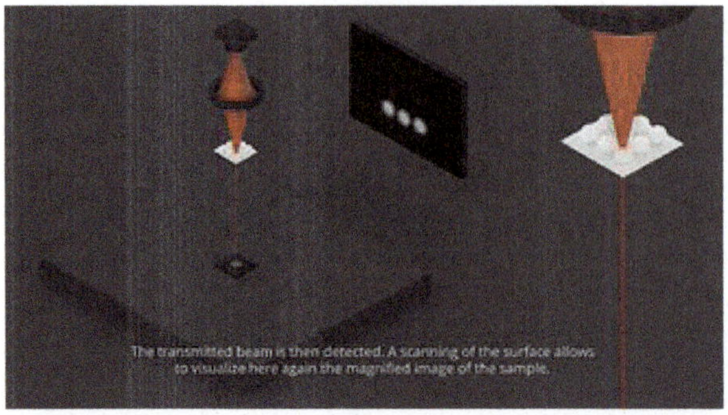

Operating principe of a Transmission Electron Microscope

The original form of electron microscope, the transmission electron microscope (TEM) uses a high voltage electron beam to illuminate the specimen and create an image. The electron beam is produced by an electron gun, commonly fitted with a tungsten filament cathode as the electron source. The electron beam is accelerated by an anode typically at +100 keV (40 to 400 keV) with respect to the cathode, focused by electrostatic and electromagnetic lenses, and transmitted through the specimen that is in part transparent to electrons and in part scatters them out of the beam. When it emerges from the specimen, the electron beam carries information about the structure of the specimen that is magnified by the objective lens system of the microscope. The spatial variation in this information (the "image") may be viewed by projecting the magnified electron image onto a fluorescent viewing screen coated with a phosphor or scintillator material such as zinc sulfide. Alternatively, the image can be photographically recorded by exposing a photographic film or plate directly to the electron beam, or a high-resolution phosphor may be coupled by means of a lens optical system or a fibre optic light-guide to the sensor of a digital camera. The image detected by the digital camera may be displayed on a monitor or computer.

The resolution of TEMs is limited primarily by spherical aberration, but a new generation of aberration correctors have been able to partially overcome spherical aberration to increase resolution. Hardware correction of spherical aberration for the high-resolution transmission electron microscopy (HRTEM) has allowed the production of images with resolution below 0.5 angstrom (50 picometres) and magnifications above 50 million times. The ability to determine the positions of atoms within materials has made the HRTEM an important tool for nano-technologies research and development.

Transmission electron microscopes are often used in electron diffraction mode. The advantages of electron diffraction over X-ray crystallography are that the specimen need not be a single crystal or even a polycrystalline powder, and also that the Fourier transform reconstruction of the object's magnified structure occurs physically and thus avoids the need for solving the phase problem faced by the X-ray crystallographers after obtaining their X-ray diffraction patterns of a single crystal or polycrystalline powder.

One major disadvantage of the transmission electron microscope is the need for extremely thin sections of the specimens, typically about 100 nanometers. Creating these thin sections for biological and materials specimens is technically very challenging. Semiconductor thin sections can be made using a focused ion beam. Biological tissue specimens are chemically fixed, dehydrated and embedded in a polymer resin to stabilize them sufficiently to allow ultrathin sectioning. Sections of biological specimens, organic polymers and similar materials may require staining with heavy atom labels in order to achieve the required image contrast.

Scanning Electron Microscope (SEM)

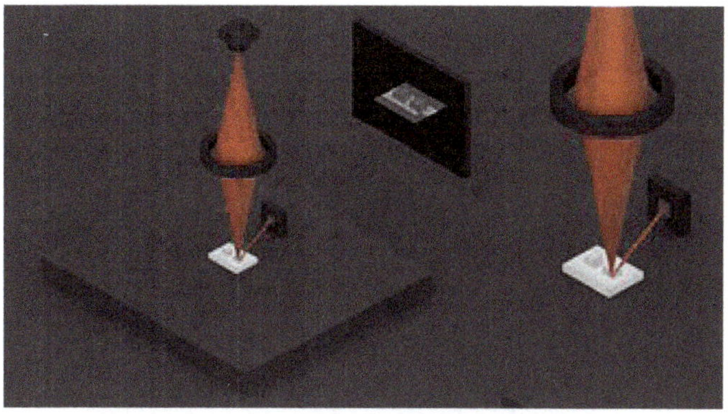

Operating principe of a Scanning Electron Microscope

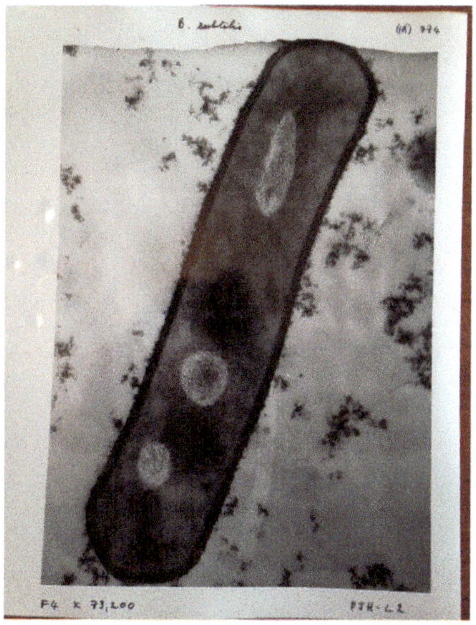

Image of bacillus subtilis taken with a 1960s electron microscope.

The SEM produces images by probing the specimen with a focused electron beam that is scanned across a rectangular area of the specimen (raster scanning). When the electron beam interacts with the specimen, it loses energy by a variety of mechanisms. The lost energy is converted into alternative forms such as heat, emission of low-energy secondary electrons and high-energy back-scattered electrons, light emission (cathodoluminescence) or X-ray emission, all of which provide signals carrying information about the properties of the specimen surface, such as its topography and composition. The image displayed by an SEM maps the varying intensity of any of these signals into the image in a position corresponding to the position of the beam on the specimen when the signal was generated. In the SEM image of an ant shown below and to the right, the image was constructed from signals produced by a secondary electron detector, the normal or conventional imaging mode in most SEMs.

Generally, the image resolution of an SEM is lower than that of a TEM. However, because the SEM images the surface of a sample rather than its interior, the elecrons do not have to travel through the sample. This reduces the need for extensive sample preparation to thin the specimen to electron transparency. The SEM is able to image bulk samples that can fit on its stage and still be maneuvered, including a height less than the working distance being used, often 4 milimeters for high resolution images. The SEM also has a great depth of field, and so can produce images that are good representations of the three-dimensional surface shape of the sample. Another advantage of SEMs comes with environmental scanning electron microscopes (ESEM) that can produce images of good quality and resolution with hydrated samples or in low, rather than high, vacuum or under chamber gases. This facilitates imaging unfixed biological samples that are unstable in the high vacuum of conventional electron microscopes.

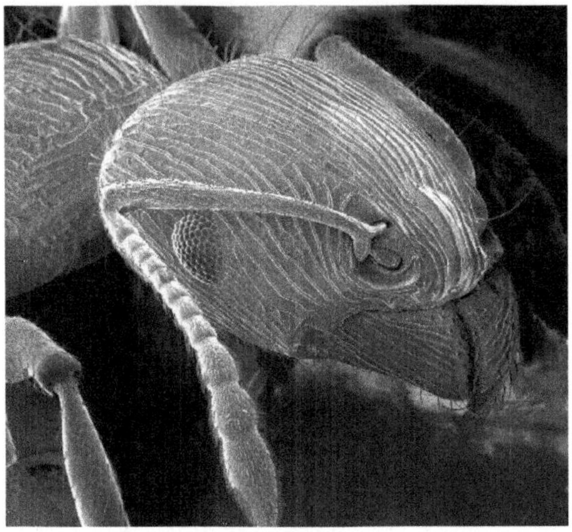

An image of an ant in a scanning electron microscope

Color

In their most common configurations, electron microscopes produce images with a single brightness value per pixel, with the results usually rendered in grayscale. However, often these images are then colorized through the use of feature-detection software, or simply by hand-editing using a graphics editor. This may be done to clarify structure or for aesthetic effect and generally does not add new information about the specimen.

In some configurations information about several specimen properties is gathered per pixel, usually by the use of multiple detectors. In SEM, the attributes of topography and material contrast can be obtained by a pair of backscattered electron detectors and such attributes can be superimposed in a single color image by assigning a different primary color to each attribute. Similarly, a combination of backscattered and secondary electron signals can be assigned to different colors and superimposed on a single color micrograph displaying simultaneously the properties of the specimen.

Some types of detectors used in SEM have analytical capabilities, and can provide several items of data at each pixel. Examples are the Energy-dispersive X-ray spectroscopy (EDS) detectors used in elemental analysis and Cathodoluminescence microscope (CL) systems that analyse the intensity and spectrum of electron-induced luminescence in (for example) geological specimens. In SEM systems using these detectors it is common to color code the signals and superimpose them in a single color image, so that differences in the distribution of the various components of the specimen can be seen clearly and compared. Optionally, the standard secondary electron image can be merged with the one or more compositional channels, so that the specimen's structure and composition can be compared. Such images can be made while maintaining the full integrity of the original signal, which is not modified in any way.

Reflection Electron Microscope (REM)

In the reflection electron microscope (REM) as in the TEM, an electron beam is incident on a surface but instead of using the transmission (TEM) or secondary electrons (SEM), the reflected beam of elastically scattered electrons is detected. This technique is typically coupled with reflection high energy electron diffraction (RHEED) and *reflection high-energy loss spectroscopy (RHELS)*. Another variation is spin-polarized low-energy electron microscopy (SPLEEM), which is used for looking at the microstructure of magnetic domains.

Scanning Transmission Electron Microscope (STEM)

The STEM rasters a focused incident probe across a specimen that (as with the TEM) has been thinned to facilitate detection of electrons scattered *through* the specimen. The high resolution of the TEM is thus possible in STEM. The focusing action (and aberrations) occur before the electrons hit the specimen in the STEM, but afterward in the TEM. The STEMs use of SEM-like beam rastering simplifies annular dark-field imaging, and other analytical techniques, but also means that image data is acquired in serial rather than in parallel fashion. Often TEM can be equipped with the scanning option and then it can function both as TEM and STEM.

Sample Preparation

An insect coated in gold for viewing with a scanning electron microscope

Materials to be viewed under an electron microscope may require processing to produce a suitable sample. The technique required varies depending on the specimen and the analysis required:

- *Chemical fixation* – for biological specimens aims to stabilize the specimen's mobile macromolecular structure by chemical crosslinking of proteins with aldehydes such as formaldehyde and glutaraldehyde, and lipids with osmium tetroxide.

- *Negative stain* – suspensions containing nanoparticles or fine biological material (such as viruses and bacteria) are briefly mixed with a dilute solution of an electron-opaque solution such as ammonium molybdate, uranyl acetate (or formate), or phosphotungstic acid. This mixture is applied to a suitably coated EM grid, blotted, then allowed to dry. Viewing of this preparation in the TEM should be carried out without delay for best results. The method is important in microbiology for fast but crude morphological identification, but can also be used as the basis for high resolution 3D reconstruction using EM tomography methodology when carbon films are used for support. Negative staining is also used for observation of nanoparticles.

- *Cryofixation* – freezing a specimen so rapidly, in liquid ethane, and maintained at liquid nitrogen or even liquid helium temperatures, so that the water forms vitreous (non-crystalline) ice. This preserves the specimen in a snapshot of its solution state. An entire field called cryo-electron microscopy has branched from this technique. With the development of cryo-electron microscopy of vitreous sections (CEMOVIS), it is now possible to observe samples from virtually any biological specimen close to its native state.

- *Dehydration* – or replacement of water with organic solvents such as ethanol or acetone, followed by critical point drying or infiltration with embedding resins. Also freeze drying.

- *Embedding, biological specimens* – after dehydration, tissue for observation in the transmission electron microscope is embedded so it can be sectioned ready for viewing. To do this the tissue is passed through a 'transition solvent' such as propylene oxide (epoxypropane) or acetone and then infiltrated with an epoxy resin such as Araldite, Epon, or Durcupan; tissues may also be embedded directly in water-miscible acrylic resin. After the resin has been polymerized (hardened) the sample is thin sectioned (ultrathin sections) and stained – it is then ready for viewing.

- *Embedding, materials* – after embedding in resin, the specimen is usually ground and polished to a mirror-like finish using ultra-fine abrasives. The polishing process must be performed carefully to minimize scratches and other polishing artifacts that reduce image quality.

- *Metal shadowing* – Metal (e.g. platinum) is evaporated from an overhead electrode and applied to the surface of a biological sample at an angle. The surface topography results in variations in the thickness of the metal that are seen as variations in brightness and contrast in the electron microscope image.

- *Replication* – A surface shadowed with metal (e.g. platinum, or a mixture of carbon and platinum) at an angle is coated with pure carbon evaporated from carbon electrodes at right angles to the surface. This is followed by removal of the specimen material (e.g. in an acid bath, using enzymes or by mechanical separation) to produce a surface replica that records the surface ultrastructure and can be examined using transmission electron microscopy.

- *Sectioning* – produces thin slices of specimen, semitransparent to electrons. These can be cut on an ultramicrotome with a diamond knife to produce ultra-thin sections about 60–90 nm thick. Disposable glass knives are also used because they can be made in the lab and are much cheaper.

- *Staining* – uses heavy metals such as lead, uranium or tungsten to scatter imaging electrons and thus give contrast between different structures, since many (especially biological) materials are nearly "transparent" to electrons (weak phase objects). In biology, specimens can be stained "en bloc" before embedding and also later after sectioning. Typically thin sections are stained for several minutes with an aqueous or alcoholic solution of uranyl acetate followed by aqueous lead citrate.
- *Freeze-fracture or freeze-etch* – a preparation method particularly useful for examining lipid membranes and their incorporated proteins in "face on" view. The fresh tissue or cell suspension is frozen rapidly (cryofixation), then fractured by breaking or by using a microtome while maintained at liquid nitrogen temperature. The cold fractured surface (sometimes "etched" by increasing the temperature to about −100 °C for several minutes to let some ice sublime) is then shadowed with evaporated platinum or gold at an average angle of 45° in a high vacuum evaporator. A second coat of carbon, evaporated perpendicular to the average surface plane is often performed to improve stability of the replica coating. The specimen is returned to room temperature and pressure, then the extremely fragile "pre-shadowed" metal replica of the fracture surface is released from the underlying biological material by careful chemical digestion with acids, hypochlorite solution or SDS detergent. The still-floating replica is thoroughly washed free from residual chemicals, carefully fished up on fine grids, dried then viewed in the TEM.
- *Ion beam milling* – thins samples until they are transparent to electrons by firing ions (typically argon) at the surface from an angle and sputtering material from the surface. A subclass of this is focused ion beam milling, where gallium ions are used to produce an electron transparent membrane in a specific region of the sample, for example through a device within a microprocessor. Ion beam milling may also be used for cross-section polishing prior to SEM analysis of materials that are difficult to prepare using mechanical polishing.
- *Conductive coating* – an ultrathin coating of electrically conducting material, deposited either by high vacuum evaporation or by low vacuum sputter coating of the sample. This is done to prevent the accumulation of static electric fields at the specimen due to the electron irradiation required during imaging. The coating materials include gold, gold/palladium, platinum, tungsten, graphite, etc.
- *Earthing* – to avoid electrical charge accumulation on a conductive coated sample, it is usually electrically connected to the metal sample holder. Often an electrically conductive adhesive is used for this purpose.

Disadvantages

Electron microscopes are expensive to build and maintain, but the capital and running costs of confocal light microscope systems now overlaps with those of basic electron microscopes. Microscopes designed to achieve high resolutions must be housed in stable buildings (sometimes underground) with special services such as magnetic field cancelling systems.

The samples largely have to be viewed in vacuum, as the molecules that make up air would scatter the electrons. An exception is the environmental scanning electron microscope, which allows hydrated samples to be viewed in a low-pressure (up to 20 Torr or 2.7 kPa) and/or wet environment.

Scanning electron microscopes operating in conventional high-vacuum mode usually image conductive specimens; therefore non-conductive materials require conductive coating (gold/palladium alloy, carbon, osmium, etc.). The low-voltage mode of modern microscopes makes possible the observation of non-conductive specimens without coating. Non-conductive materials can be imaged also by a variable pressure (or environmental) scanning electron microscope.

Small, stable specimens such as carbon nanotubes, diatom frustules and small mineral crystals (asbestos fibres, for example) require no special treatment before being examined in the electron microscope. Samples of hydrated materials, including almost all biological specimens have to be prepared in various ways to stabilize them, reduce their thickness (ultrathin sectioning) and increase their electron optical contrast (staining). These processes may result in *artifacts*, but these can usually be identified by comparing the results obtained by using radically different specimen preparation methods. Since the 1980s, analysis of cryofixed, vitrified specimens has also become increasingly used by scientists, further confirming the validity of this technique.

Concept of Resolution

- In light microscopes, we mostly use sources with the visible spectrum ranging between 400 – 700 nm. For example, taking green light with λ = 550 nm as an example, we can obtain the resolution of 300 nm (according to Rayleigh's criterion and discussed earlier).
- But, the distance between two atoms in solids is around 0.2 nm. Hence, we need to utilize the wavelength below the visible region.
- Based on wave-particle duality, we know that electron has wave-like properties. The relation between the electron accelerating voltage (Vo) and wavelength is given as:

$$\lambda = \frac{h}{p} = \frac{h}{\sqrt{2m_0 e v_0 \left(1 + \frac{e v_0}{2 m_0 C^2}\right)}}$$

- The above equation can further be modified to The Energy – Lambda relation →

$$\lambda(\mathring{A}) = \frac{12.398}{\sqrt{kV * (1022 + kV)}}$$

Substituting different values of operating voltage, we can generate the table of resulting wavelength of electron.

kV	Pico meter
100	3.701
160	2.851
200	2.508
300	1.969
400	1.644
1000	0.872

Electron Microscope: BASICs

e⁻ scattering and e⁻ - atom interaction: Basics and Principles

Electron Scattering:

An electron passing through a solid may be scattered

(iii) Not at all

(iv) Only once (single scattering)

(v) Several times (plural scattering), or

(vi) Very many times (multiple scatterings)

- Each scattering event might be elastic (involves no loss of energy) or inelastic (involves loss of energy).
- The scattered electron is most likely to be forward scattered but there is a considerable chance for back scattering.
- The probability of scattering is described in terms of either an "interaction cross-section" or a mean free path.
- When the solid specimen is thicker (twice the mean free path), plural scattering is likely to be happened.

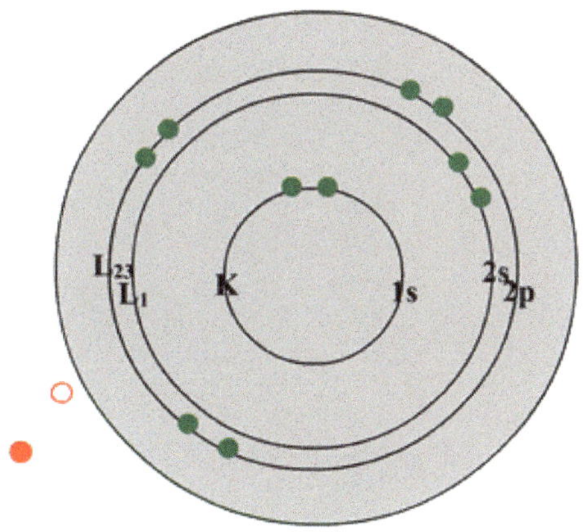

Schematic of excitation of single electron in a material.

Electron-atom Interaction

When a thin specimen is illuminated with primary electrons, there are four main processes by which a high energy electron can lose energy to an atom.

- The excitation of a single electron from the valence band (open circle)

The primary electron (marked as solid red color) strikes the sea of valence electrons and knocks out one electron (open circle) from the sea. This process is called as the excitation of a single electron from the valence band.

Electron-Atom Interaction

- The excitation of a plasmon

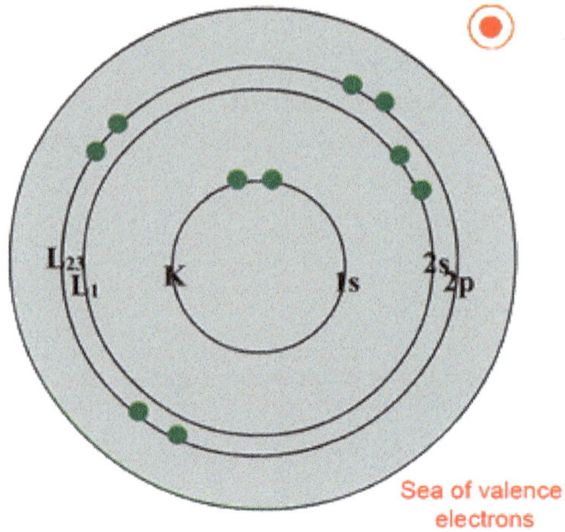

Schematic of excitation of a plasmon.

The primary electron (marked as solid red color) strikes the sea of valence electrons and moves away without removing any electrons from the sea. This process is called as the excitation of a plasmon.

- The excitation of an inner shell electron (from the K or L shell)

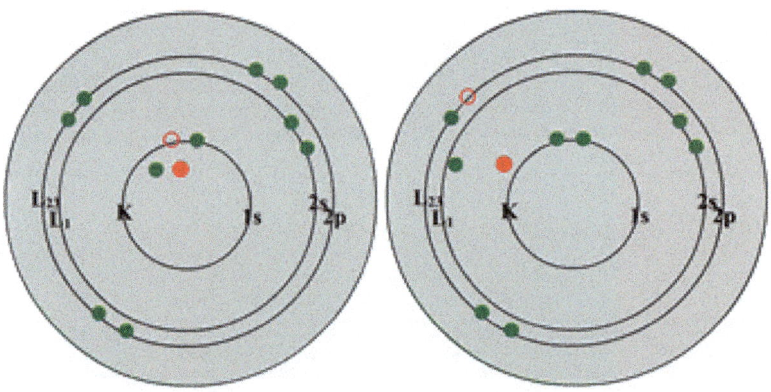

Schematic of excitation of a plasmon.

Let us consider that the primary electron (marked as solid red color) strikes the electrons in the inner shells (K or L). This results in a removal of electron from the inner shells resulting in an unstable conditions. Subsequently, the relaxation of inner shell excited atoms may occur by the emission of an X-ray or an Auger electron as demonstrated in the following slides.

Basics and Principles

Characteristic X-ray Emission

- After an inner shell excitation, an atom has an energy above its ground state. It can relax and lose some of this energy in several (two) ways as shown in figure. The outer shell electrons jump into inner shell to fill the vacancy in the inner shell.

- The energy difference due to the filling of the vacancy is given off as a single X-ray photon characterizing the atom. This is the basic of the elemental analysis using Energy Dispersive X-ray Spectroscopy (EDS).

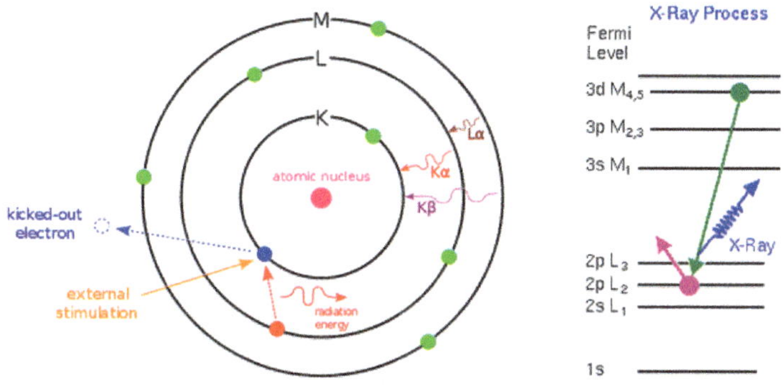

Schematic of X-ray emission during electron-atom interaction.

Auger Electron Emission

- After an inner shell excitation, an atom has an energy above its ground state. It can relax and lose some of this energy in several (two) ways as shown in figure. The outer shell electrons jump in to inner shell to fill the vacancy in the inner shell.
- The energy difference due to the filling of the vacancy is given off as leaving of one of the outer electrons (pale green circle in the left side figure and red circle in right side figure). It carries a characteristic kinetic energy. This technique is used to study the surface features of the samples.

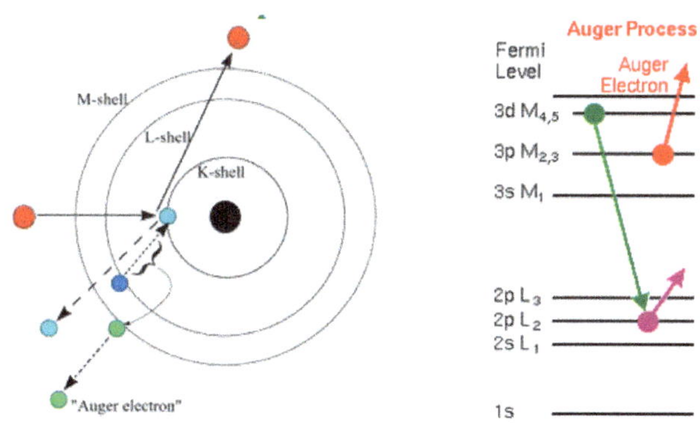

Schematic of Auger electron emission during electron-atom interaction.

References

- Appendix A, Journal of the Royal Microscopical Society, 1906; page 716. A discussion of Zeiss measuring microscopes
- Ritter, Karl; Rising, Malin (8 October 2014). "2 Americans, 1 German win chemistry Nobel". AP News. Retrieved 8 October 2014
- Manuel Gunkel; et al. (2009). "Dual color localization microscopy of cellular nanostructures". Biotechnology journal. 4 (6): 927–38. doi:10.1002/biot.200900005. PMID 19548231
- Ian M. Watt (1997). The Principles and Practice of Electron Microscopy (2 ed.). Cambridge University Press. p. 6. ISBN 0521434564

- Chang, Kenneth (8 October 2014). "2 Americans and a German Are Awarded Nobel Prize in Chemistry". New York Times. Retrieved 8 October 2014
- Luft, J.H. (1961). "Improvements in epoxy resin embedding methods". The Journal of biophysical and biochemical cytology. 9 (2). p. 409. PMC 2224998. PMID 13764136
- Albert Van Helden; Sven Dupré; Rob van Gent (2010). The Origins of the Telescope. Amsterdam University Press. p. 24. ISBN 978-90-6984-615-6
- JR Blueford. "Lesson 2 – Page 3, CLASSIFICATION OF MICROSCOPES". msnucleus.org. Retrieved 15 January 2017
- Juniper, B.E.; Bradley, D.E. (1958). "The carbon replica technique in the study of the ultrastructure of leaf surfaces". Journal of ultrastructure research. 2 (1): 16–27. doi:10.1016/s0022-5320(58)90045-5
- Gould, Stephen Jay (2000). "Chapter 2: The Sharp-Eyed Lynx, Outfoxed by Nature". The Lying Stones of Marrakech: Penultimate Reflections in Natural History. New York, N.Y: Harmony. ISBN 0-224-05044-3
- Kenneth, Spring; Keller, H. Ernst; Davidson, Michael W. "Microscope objectives". Olympus Microscopy Resource Center. Retrieved 29 Oct 2008
- Moor H, Mühlethaler K. FINE STRUCTURE IN FROZEN-ETCHED YEAST CELLS. The Journal of Cell Biology. 1963;17(3):609–628
- Stephen G. Lipson, Ariel Lipson, Henry Lipson, Optical Physics 4th Edition, Cambridge University Press, ISBN 978-0-521-49345-1
- "Demonstration of a Low-Cost, Single-Molecule Capable, Multimode Optical Microscope". Retrieved 25 February 2009.
- Da Silva, Pedro Pinto; Branton, Daniel (1970-06-01). "MEMBRANE SPLITTING IN FREEZE-ETCHING". The Journal of Cell Biology. 45 (3): 598–605. PMC 2107921. PMID 4918216
- Burgess, Jeremy (1987). Under the Microscope: A Hidden World Revealed. CUP Archive. p. 11. ISBN 0-521-39940-8
- Sabanay, I.; Arad, T.; Weiner, S.; Geiger, B. (1991). "Study of vitrified, unstained frozen tissue sections by cryoimmunoelectron microscopy". Journal of Cell Science. 100 (1): 227–236. PMID 1795028
- "The Scale of Things". Office of Basic Energy Sciences, U.S. Department of Energy. 2006-05-26. Archived from the original on 2010-02-01. Retrieved 2010-01-31
- Von Appen, A.; Beck, M. (2016). "Structure determination of the nuclear pore complex with three-dimensional cryo electron microscopy". Journal of Molecular Biology. 428 (10A): 2001–2010
- "SPLEEM". National Center for Electron Microscopy (NCEM). Archived from the original on 2010-05-29. Retrieved 2010-01-31

3

Microscopic Techniques used for Characterization of Materials

Different types of microscopes use different techniques in analysis. One microscopy technique, transmission electron microscope, uses electrons to achieve higher level of resolution than a light microscope. It can see objects which are sized in angstroms. The chapter also explores the mechanism of scanning electron microscope. It is a type of electron microscope that uses focused beams to produce an image. Science and technology have undergone rapid developments in the past decade which has resulted in the discovery of significant techniques in microscopy; which have been extensively detailed in this chapter.

Microscopic Techniques

Electron-Atom Interaction:

- In all the cases, the incoming primary electrons can be scattered elastically or in-elastically.
- The interactions include the excitation of inner shell electrons, the creation of plasmons and interactions with single valence electrons, the excitation of a phonon.
- We have shown the formation of various processes in the earlier slides for each incoming primary electron separately. However, the subsequent relaxation of inner shell excited atoms may occur by the emission of an X-ray or an Auger electron.

Overall Interaction Between the Electron and Specimen

Figures summarize the electron, X-ray and light signals generated by the interaction between electron beam and specimen. The component elements can be analyzed by measuring the characteristics X-rays with an energy dispersive X-ray spectroscopy (EDS). While backscattered, Auger, secondary electrons, etc., occur, a part of the electron beam pass through the specimen for its thin thickness. Electron Energy Loss Spectroscopy (EELS) allows to detect the bonding information of the specimen.

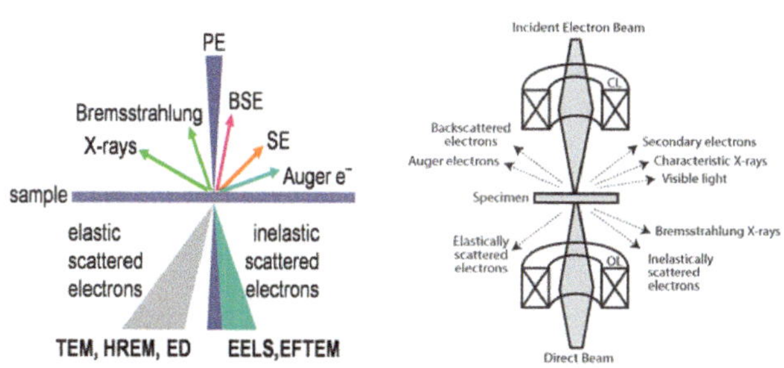

Products of electron and atom interaction.

Transmission Electron Microscope

- An instrument to study the microstructural properties of a material with atomic resolution.
- Operates on the same basic principles as a light microscope but uses electrons instead of light. Electrons have much lower wavelength and make it possible to get a resolution a thousand times better than with a light microscope. Table summarizes various features of light versus electron microscope.

FEATURE	Light Microscope	Electron Microscope
Electromagnetic spectrum used	Visible light 760 nm (red) – 390 nm Colours visible	Electrons App. < 1 nm Monochrome
Maximum resolving power	app. 200 nm	0.2 nm Fine detail
Maximum magnification	×1,000 – ×1,500	×5,00,000
Radiation source	Tungsten or quartz halogen lamp	High voltage (> 100 kV) Tungsten or LaB_6
Lenses	Glass based	Magnets based (Lorentz force)
Interior	Air-filled	High vacuum ($<10^{-7}$ Torr)
Focusing screen	Human eye or photographic film	Fluorescent screen, photographic film

This table summarizes the different features of the light and electron microscopes.

Comparision of various features between light and electron microscope.

- One can see the objects in the order of a few angstroms (10^{-10} m). In particular, very fine details of a cell or material down to near atomic levels can be probed.
- The possibility of having high magnifications in TEM has made it as a valuable tool in both medical, biological and materials research.

Figure displays the typical arrangements of slide projector and TEM technique. In TEM, the magnetic lenses are used to control the e-beam path (quite similar to the glass lenses used in the light microscope) and the final image of the specimen is projected on fluorescent screen (projection screen in slide projector). A suitable aperture is used to select the important portion of the materials for the investigation.

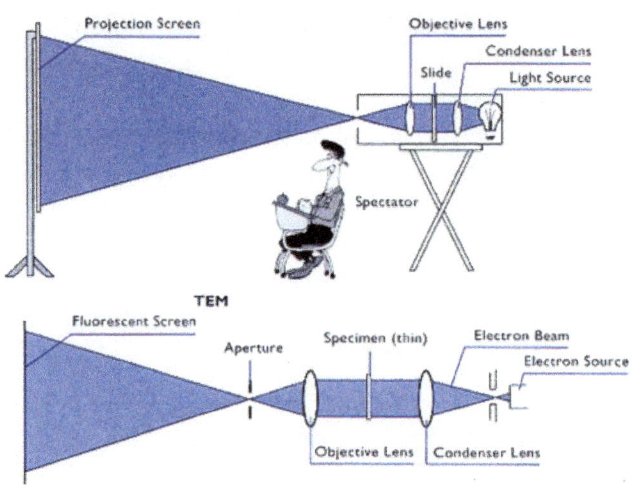

Typical arrangment of slide projector and TEM.

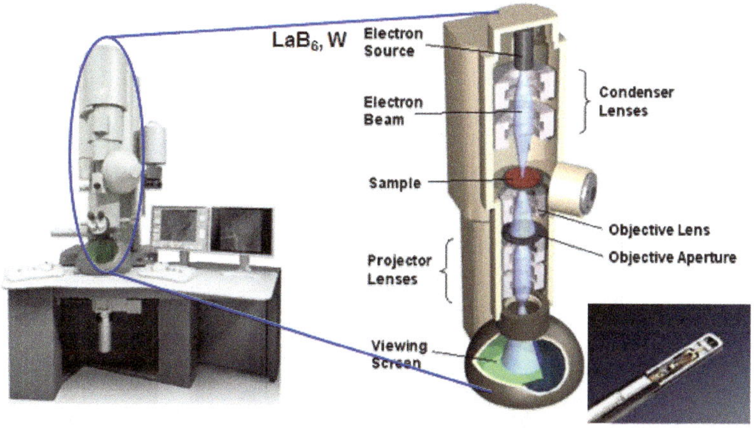

Photoview of Technai G2 F30 machine and interview column view of the system.

Figure shows the photographic view of Technai G2 F30 machine and the internal view of the column and e-beam path diagram along with the sample holder. The e-beam generated from the gun is made to pass through the specimen and the transmitted beams are projected on the fluorescent screen with the help of magnetic lenses.

Instrumental Part

Sources (Electron Guns): Thermionic Guns

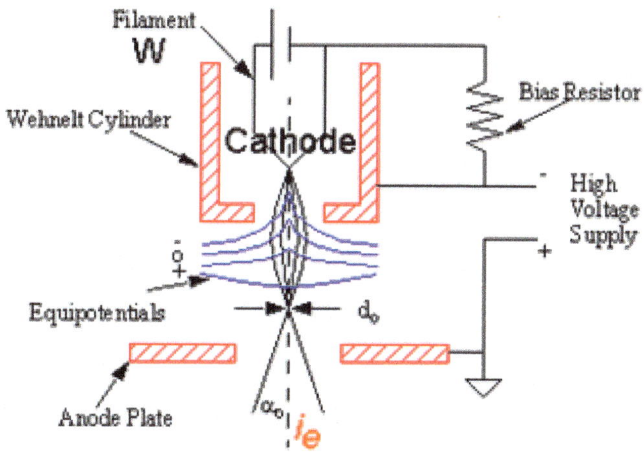

Schematic of electron gun in TEM.

- 100 μm diameter tungsten cathode or LaB6 cathode is used in the form of fine wire in hairpin shape (V-shape).
- Passing high current generates a temperature of nearly 2400°C.
- The electrons potential distribution of 0 to 2 volts. With a bias voltage (0 to 500V), electrons can be accelerated toward the anode. The typical lifetime is ~ 50 hours.
- A high voltage is placed between the cathode and the anode, modified by a potential on the Wehnelt which acts as the grid in a triode system. The electric field from the Wehnelt focuses the electrons into a crossover.

Lenses (Condenser):

Electron lenses are the magnetic equivalent of the glass lenses used in the light microscope. The behaviour of all the lenses in a TEM can be compared to the action of a convex (converging) glass lens on monochromatic light.

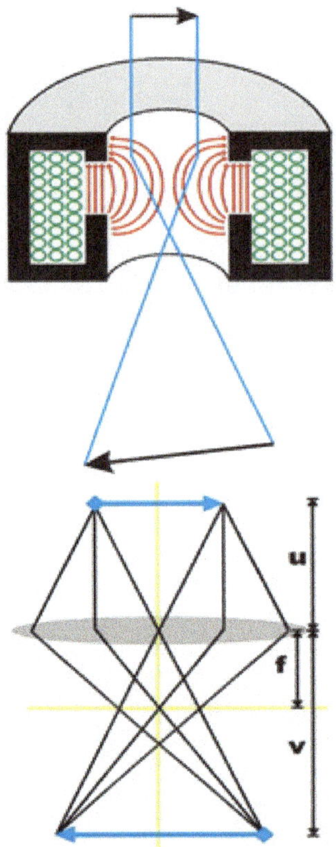

Schematic of condenser lens in TEM and its Light Optical Analogue.

A strong magnetic field due to the current carrying coils acts as a convex lens and bringing off axis rays back to focus.

In a magnetic field, an electron experiences the Lorentz force: $F = -e (E + v \times B)$

E: strength of electric field, B: strength of magnetic field, e/v: charge/velocity of electrons

The focusing effect of a magnetic lens therefore increases with the magnetic field B (by controlling i). Focal length can be altered by changing the strength of the current.

Second condenser lens affects the convergence of the beam at the specimen and the diameter of the illuminated area of the specimen.

Objective Lens:

- The objective lens forms an inverted initial image, which is subsequently magnified. In addition, in the back focal plane of the objective lens a diffraction pattern is formed. Hence, the objective aperture has to be utilized to select the components. Note that the objective lens usually would not provide a magnification of more than 50*X* and any magnification of the image is mainly done by using projector lenses.

Projector Lens:

- Magnification in the electron microscope can be varied from few hundreds to several hundred thousands of times.
- This is achieved by varying the strength of the projector lenses and intermediate lenses as shown in Figure.

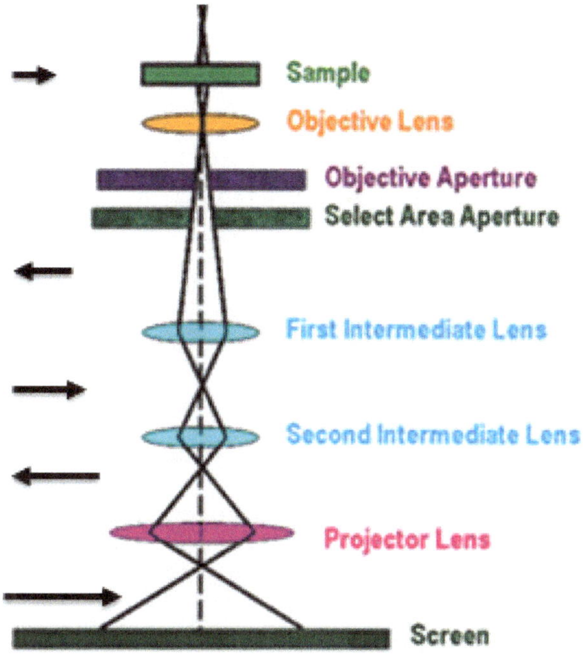

Schematic of lens system including projector lens in TEM.

- Note that all lenses need not (unnecessarily) be used at lower magnifications.
- The introduction of each lens in the system for magnification changes the direction of final image on a fluorescent screen.

TEM

TEM: Sample Preparation:

- TEM specimen should be electron transparent, uniformly thin and stable enough to withstand the transmission of electrons. As the samples are too thin, they are generally mounted on grids with and/or without mesh.
- There are many ways of sample preparation. The method chosen will depend on the type of materials (Materials science and Biological science) and the information (Nanostructure, cross-sectional) one need to obtain.
- Most important aspect to bear in mind is that the preparation technique must never affect what one observes or measures.

TEM: Sample Preparation (Thin Film)

Steps involved:

Cutting, grinding and polishing

- Cut the specimen of required size (5 mm X 5 mm) as shown in figure, which fits into the Gatan disc grinder, using ultrasonic disk-cutter. It employs a lead zirconate / titanate piezoelectric crystal to vibrate a tubular tool, pressed against the specimen, at a frequency of ~26 kHz.

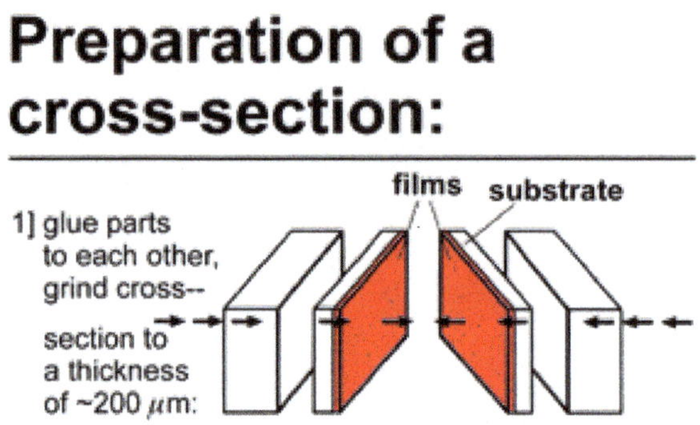

Schematic of cutting the specimen for polishing.

- One of the cut pieces of the specimen should be glued, using the low melting point wax supplied, on the specimen disc, which would be eventually mounted inside the Gatan disc grinder. This is needed for plane view imaging. In the case of cross-sectional imaging, the film portion must be glued and mount the sample across the cross sectional point of view.

Photoview of Gatan disc polisher.

- In the case of cross sectional observation, the process is considerably complicated. At first, the specimen is cut into thin slices of 1-2 mm width and 5 mm length, and then two slices of the specimen are pasted together using the glue so that the films face each other. The glue is generally made of a mixer of hardener and resin, and applied gently on the surface to be fixed with the metal piece.

- The glued specimen is now kept on a heater and heated at 120°C to bind the slices rigidly. After fixing, the glued specimen is loaded on a specimen disc and is glued using low melting point wax again.

- The specimen is continuously polished by placing the disc grinder on SiC paper with required roughness, which is fixed on a rotating polishing machine.
- The grinding thickness can be controlled systematically by adjusting the control knob of the grinder with the step size of 10 μm until the specimen is flush with the polishing surface.
- This process reduces the specimen thickness from the substrate side to around 20 μm. Care must be taken to thin the samples slowly without breaking into pieces.
- Finally, the specimen disc along with the thinned specimen is removed from the Gatan disc grinder and is placed on the heater. Upon melting the glue, the thin specimen comes out of the specimen disc.
- As the polished films have the thickness in the range of < 30 μm, it has to be mounted on a stable platform for further thinning of the films without any damages and cracks.
- Thinned films are mostly mounted on molybdenum TEM grids as shown in figure.

Schematic of loading polished sample on the grid.

- The glue made out of a mixer of hardener and resin is applied on top of the grid selectively (mid figure) followed by positioning the thin specimen on top of the glue (right one).

The glued specimen is now kept on a heater and heated at 120°C to bind the thin specimen on the grid rigidly.

- The thinned film on the grid is loaded into a low angle ion polishing machine for creating high-quality TEM specimens with large electron transparent areas.

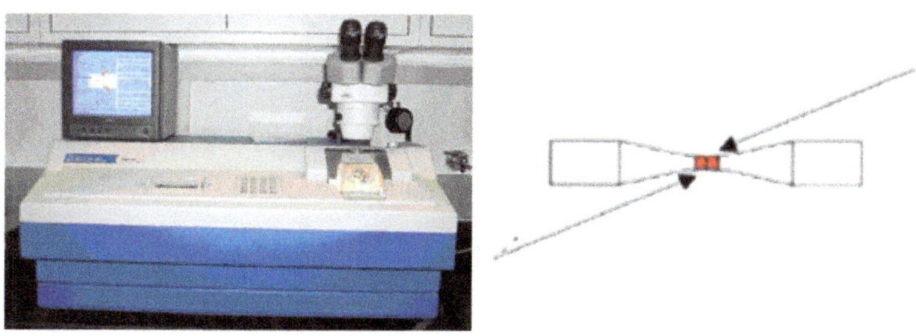

Photoview of fischione ion mill and schematic of milling process.

Ion Polishing (For Plane-view Observation)

- Two individually controlled hollow anode discharge (HAD) ion sources are focused on both specimen surfaces as shown in figure. This source is small and requires only about 0.4 sccm of gas, but produces a wide range of ion beam energies.

- The formation of high-quality TEM specimens with large electron transparent areas are schematically shown in figure (left).
- As the milling time increases, the formation of a fine hole with the electron transparent appears after certain period. On further increasing the milling time, the specimen exhibits large crack (right image of figure) and the electron transparent areas would be diminished. The time required to form uniform area varies with the samples.

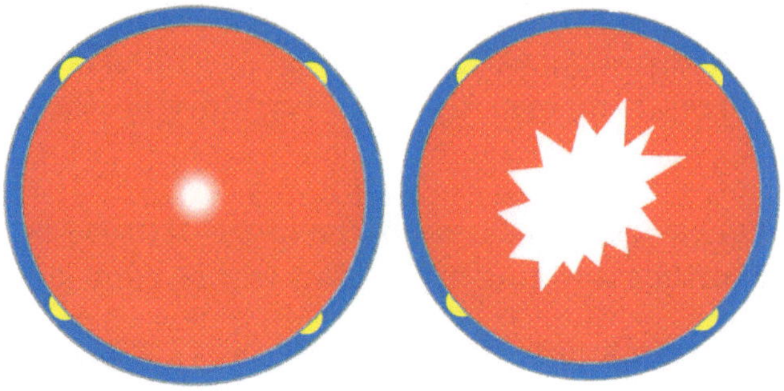

Formation of electron transparent area with milling.

Ion polishing (for cross-sectional observation):

The thinned film on the grid is loaded Precision Ion Polishing System (PIPS) for creating high-quality TEM specimens.

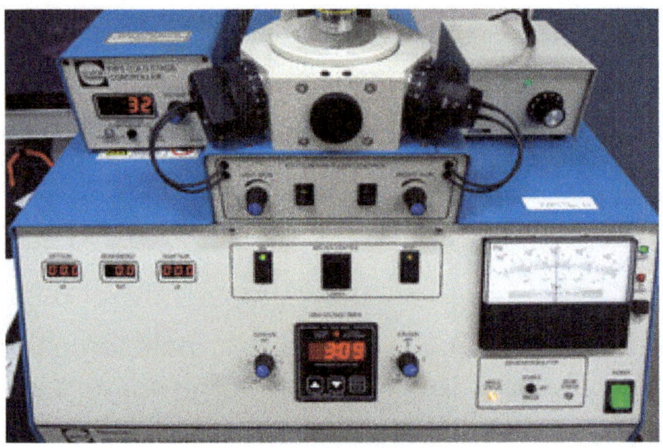

Photoview of PIPS.

- Two Penning-type ion guns (with input power of ~ 2 W at 5 kV accel. voltage) fixed on fully adjustable flanges with the design as similar to the one shown in figure.
- The sample is fixed on a holder inside the stage which can be rotated, rocked and tilted (0 to 30). For small angles, there is no interference of the ion beam with the holder or the stage.
- The possibility of etching (less than 5°) reduces differential sputtering and thermal stress.
- Redeposition of sputtered material is avoided due to the special design of the holder.

Figure displays the typical sample arm for TEM. The sample arm consists of spacing for grid loading, suitable O-ring for vacuum and handle for loading/unloading the holder into/out of TEM. The loading process of grid in the sample arm is shown schematically.

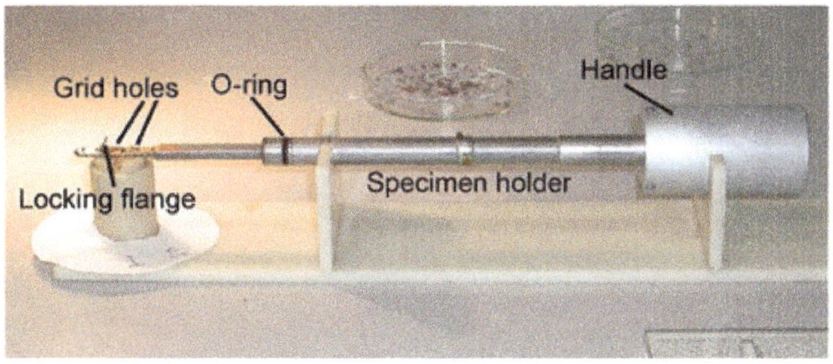

Typical sample hoder for TEM.

Image view of 915 double tilt liquid nitrogen cryo transfer holder combines both analytical and cryo transfer capabilities.

Figure displays the 915 double tilt liquid nitrogen cryo transfer holder combines both analytical and cryo transfer capabilities. This is useful to observe the structure of frozen hydrated crystalline specimens, reduce unwanted thermal effects in various analytical techniques and to facilitate EDS and EELS analyses by reducing contamination and mass loss.

TEM grid is loaded first on the sample arm. Grid is safely guarded with the spacer and the screw which provides robust operation with no mechanical binding of the specimen cradle at tilt limits.

Tilt ranges and compatibility of specimen holders vary according to the TEM manufacturer, model, pole piece gap, and the presence of in-gap accessories. Please contact your local Gatan representative for more information.

TEM: Imaging Methods:

- We shall briefly cover the different imaging modes to understand the microstructure of the sample. For simple analysis, there are four different imaging modes: (i) Bright-field imaging, (ii) Dark-field imaging, (iii) Diffraction mode and (iv) High-resolution mode.
- In TEM, imaging methods utilize the information contained in the electron waves exiting from the specimen to form an image. The projector lenses allow for the correct positioning of this electron wave distribution onto the viewing system. The observed intensity of the image (I) can be approximated as proportional to the time-average amplitude of the electron wave functions.
- Different imaging methods, therefore, attempt to modify the electron waves exiting the sample in a form that is useful to obtain information with regards to the sample, or beam itself.

- To improve the contrast in the image, the TEM may be operated at a slight defocus to enhance contrast, owing to convolution by the contrast transfer function of the TEM, which would normally decrease contrast if the sample was not a weak phase object.

Bright-field imaging mode:

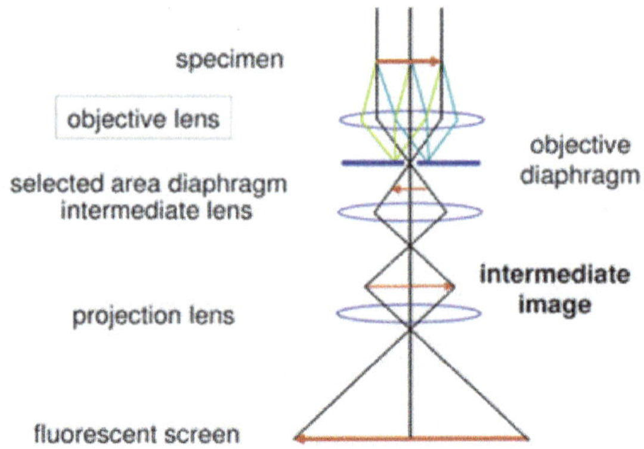

Schematic ray diagram of bright-field imaging mode.

The most common mode of operation for a TEM is the bright field imaging mode. In this mode the contrast formation is formed directly by occlusion and absorption of electrons in the sample. Thicker regions of the sample, or regions with a higher atomic number will appear dark, whilst regions with no sample in the beam path will appear bright – hence the term "bright-field".

This mode is obtained by intentionally excluding all diffracted beams and only allowing the central beam passing through the specimen as shown in the figure.

Bright-field Imaging Mode

This method is quite useful to study the microstructure of the samples in plane-view and number of multilayers from cross sectional view. The plane view images help in understanding the formation of interior microstructure and determining the average size of the particles.

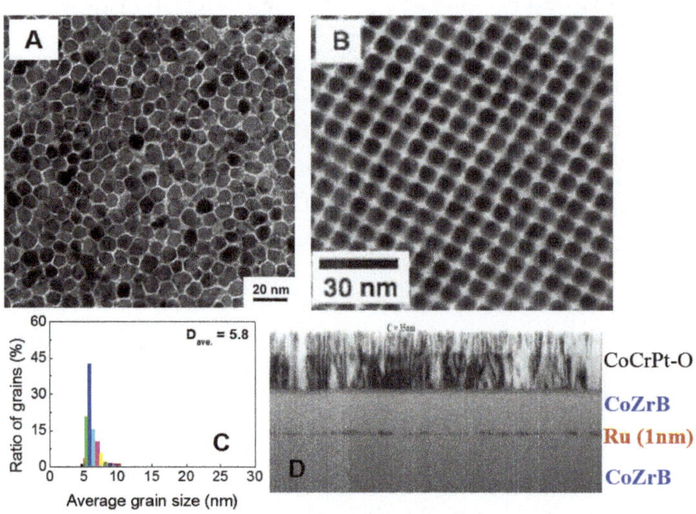

(a) TEM image of FePt-C system prepared by sputtering technique , (b) FePt particles prepared by chemical method , (c) average size of the grains determined from (a), and (d) cross-section view of the longitudinal hard disk drive.

Dark-field Imaging Mode

This is also formed by magnifying a single-beam. But this time, one of the diffracted beams is chosen by means of suitable aperture that blocks the central beam and the other diffracted beams as shown in figure.

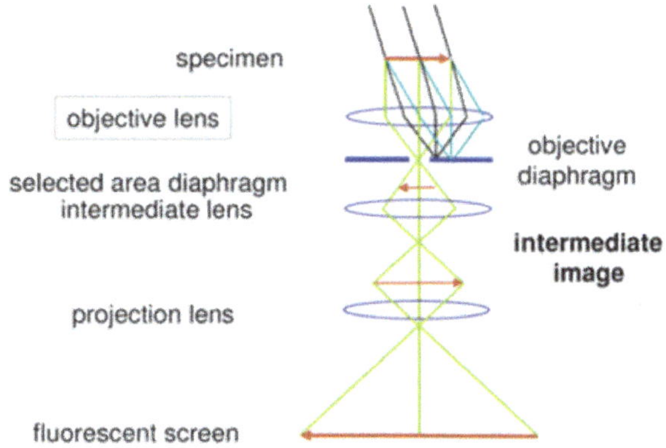

Schematic ray diagram of drak-field imaging mode.

This method is quite useful to study the orientation of the particles in a particular direction (hkl plane).

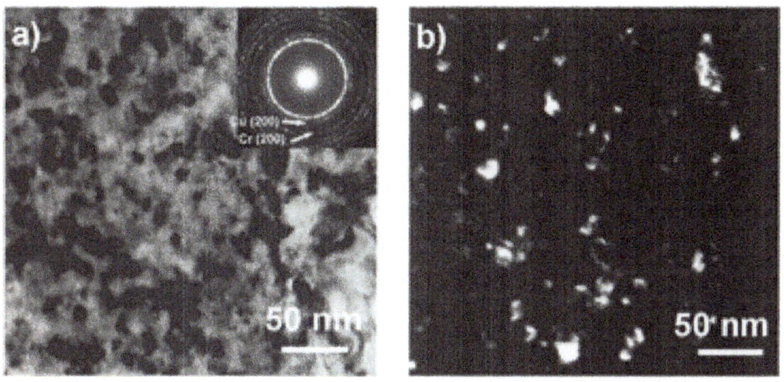

(a) TEM bright-field image and related SAED pattern of the CuCr43 composite processed by HPT up to 25 revolutions; (b) dark-field image of the same area obtained by selecting some (111)Cu and (110)Cr lattice reflections (first Debye-Scherrer ring).

Figure (a) displays the TEM bright-field image and related SAED pattern of the CuCr43 composite in nanometer range. The dark field image on the figure (b) was obtained by selecting a part of the first ring with a small aperture. This image clearly shows the grain size is in a range of only 10 to 20 nm.

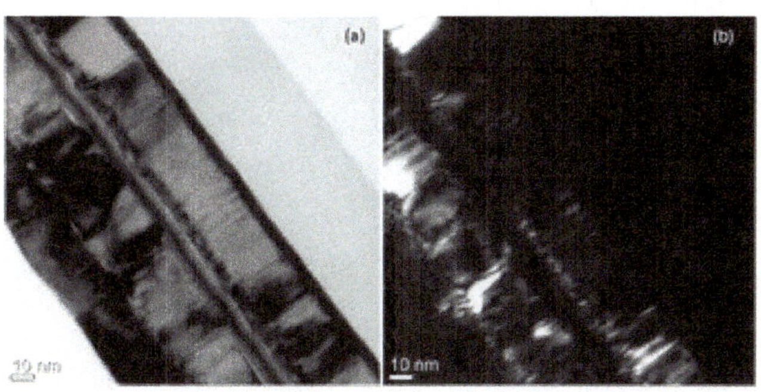

(a) TEM bright-field and (b) dark-field couple images of cross-section view of a magnetic tunneling junction shows the contrast from layer by layer.

Figure displays bright-field and dark-field images of cross-sectional view of magnetic tunnel junction with different contrast from layer by layer.

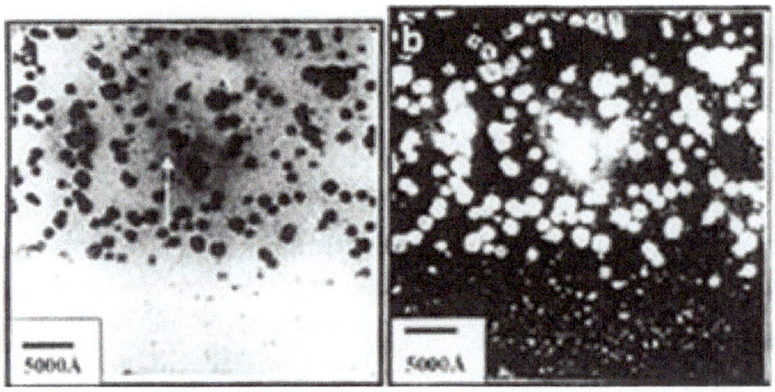

(a) Ni-Cu-2-Methoxyethanol colloid electron micrograph at 20 K magnification; 78.4% Ni and 21.6% Cu (a) Bright-field (b) Dark-field.

Figure (a) depicts zone of low contrast forming a cloud around the particles and the cluster (see the arrow) several spherical particles with an average size of 50 nm. This cloud does not allow to see us particles which can be observed in the dark field (figure b).

Diffraction Mode

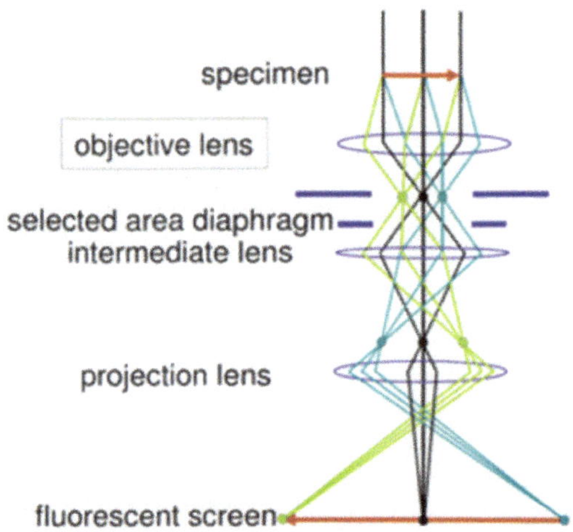

Schematic ray diagram of diffraction mode.

This is obtained by intentionally excluding central beam and collecting all diffracted beams as shown in figure below to form the diffraction pattern.

A thin crystalline specimen is subjected to a parallel beam of high-energy electrons. As the wavelength of high-energy electrons is a few thousandths of a nanometer and the spacing between atoms in a solid is about a hundred times larger, the atoms act as a diffraction grating to the electrons, which are diffracted.

As a result, the image on the screen of the TEM will be a series of spots or rings (corresponding to a satisfied diffraction condition of the sample's crystal structure).

SAED is referred to as "selected" because the user can choose part of the specimen to obtain the diffraction pattern.

As a diffraction technique, SAED can be used to identify crystal structures and examine crystal defects. It is similar to X-ray diffraction, but unique in that areas as small as several hundred nanometers in size can be examined. However, X-ray diffraction typically samples areas of several centimeters.

SAED is used primarily in material science and solid state physics. This is one of the most commonly used experimental techniques.

This is obtained by intentionally excluding central beam and collecting all diffracted beams.

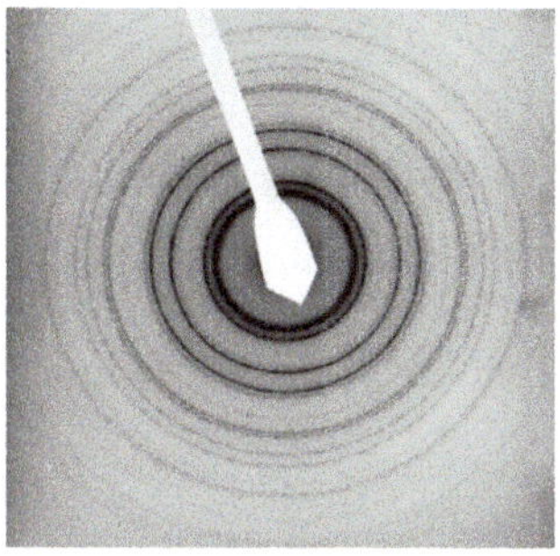

Typical diffraction pattern obtained for polycrystalline materials in TEM.

SAED Pattern Analysis for Indexing

The SAED pattern can be analyzed using the following simple relation:

$R\,d = L\,\lambda$

$L = (R.\,d)/\lambda$

where, R - Diameter of the ring, d - d - spacing, L - Camera Length, λ - Wavelength of the e -beam.

Procedure to analyze the SAED pattern:

1) Find the camera length from the calibration sample.

2) Use the same camera length to identify various SAED patterns of unknown samples.

Determining unknown parameters for Camera length calculation?

TEM: Analysis of SAED Pattern

X-ray Diffraction & SAED Pattern

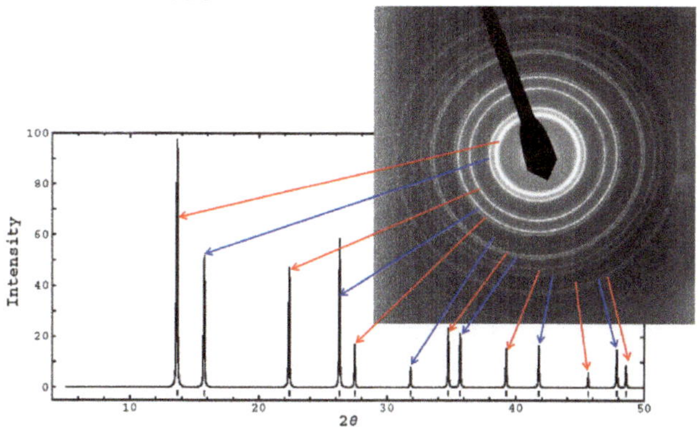

The comparison between XRD and SAED patterns.

Figure shows both XRD and SAED patterns of the gold sample. All the observed Bragg reflections in the XRD pattern are consistent with the SAED ring patterns.

X-ray Diffraction & SAED Pattern

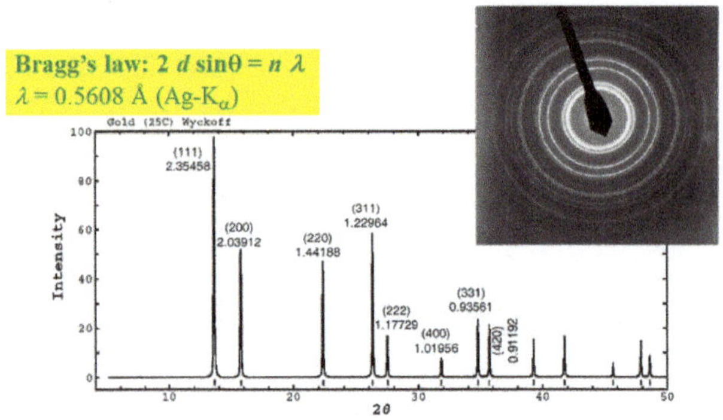

Identification of d-values from XRD.

Since the Bragg reflections, hkl planes and the lattice spacing are known for gold sample from XRD pattern, the camera length for the TEM image can be determined using L=(R.d) / λ, if one calculates the diameter of the ring using image software (photoshop) and knowing the wave-length of the electron.

Standard Sample (Gold)

Bragg's law: $2\,d\,\sin\theta = n\,\lambda$ $L = (R.d)/\lambda$

Ring #	Ring Diameter (Pixel)	d-spacing (Å)	Wavelength (λ)	Camera Length
1	4.840	2.35458	0.0164 Å (400 kV)	694.9
2	5.586	2.03912		694.5
3	7.906	1.44188		695.1
4	9.276	1.22964		695.5
5	9.675	1.17729		694.5
6	11.181	1.01956		695.1
7	12.181	0.93561		694.9
8	12.490	0.91192		694.5
9	13.694	0.83247		695.1
			Average Camera Length	694.9

As we now know the typical camera length of the TEM for a given diffraction condition with respect to the calibrated samples, the same can be used to index the planes of the unknown samples.

High-resolution TEM

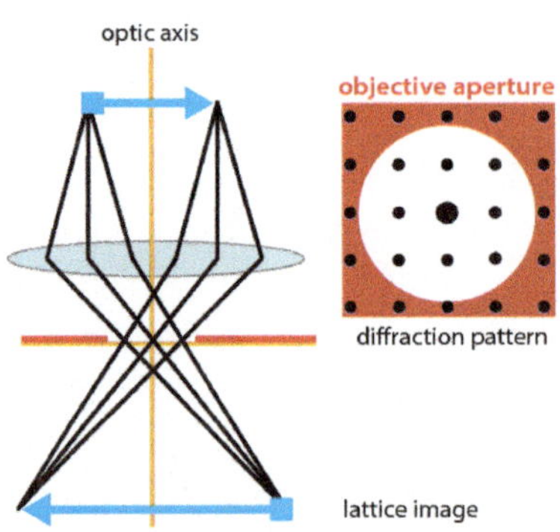

Schematic of ray diagram for HRTEM imaging mode.

High-resolution transmission electron microscopy (HRTEM) is an imaging mode of the transmission electron microscope (TEM) that allows for direct imaging of the atomic structure of the sample. HRTEM is a powerful tool to study the properties of materials on the atomic scale such as semiconductors, metals and nanoparticles.

The primary transmitted and one or more of the diffracted beams are made to recombine as shown in figure, thus preserving their phases and amplitudes. The image is formed by the interference of the diffracted beams with the direct beam (phase contrast). High resolution mode uses small effective wavelengths (λ).

High-Resolution TEM Images

HRTEM analysis helps to understand the growth of the materials, interface morphology and quality of texture, single crystalline nature of the materials, etc. figure displays the typical HRTEM images obtained using TEM.

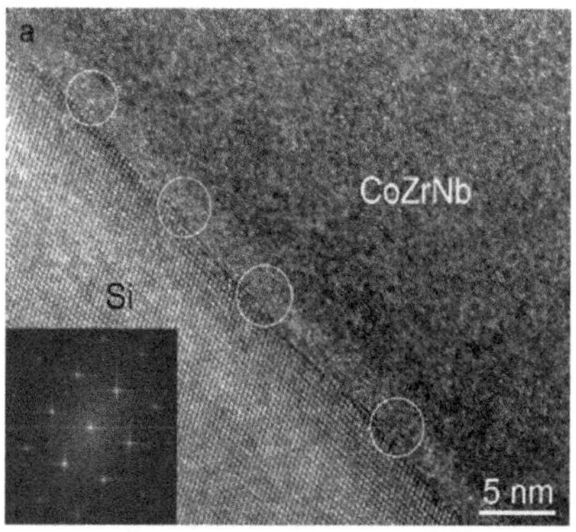

HRTEM micrograph of amorphous CoZrNb on Si substrate showing a possible inter-diffusion across the interface of about 2-3 nm.

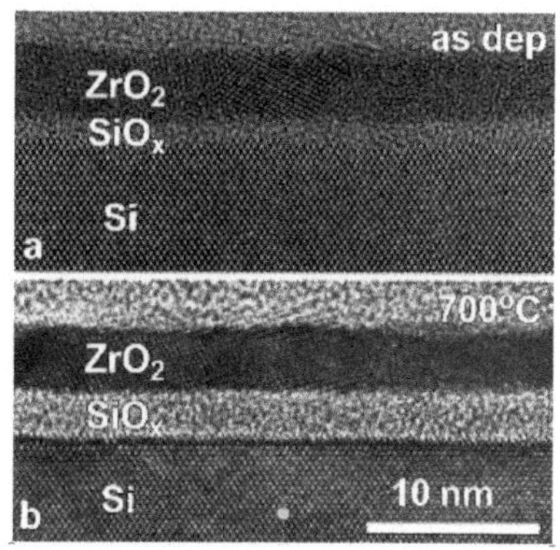

The effect of post-deposition annealing under O2 atmosphere on the microstructure of SiOx/ ZrO2 gate dielectric stacks.

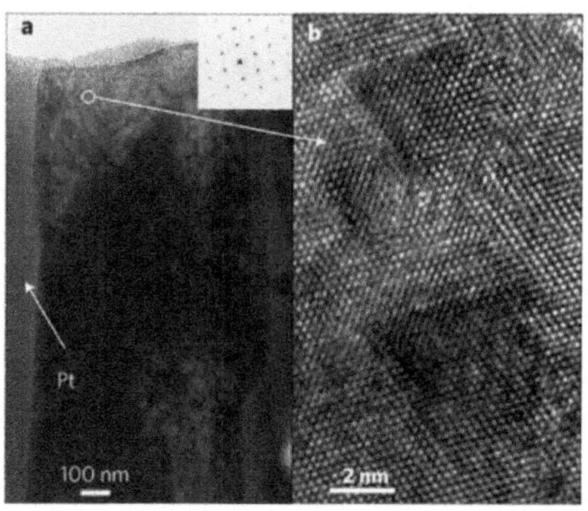

Post deformation irradiation defect characterization using high resolution TEM. (a) Low magnification TEM image of a deformed 788 nm diameter copper pillar after FIB lift out and subsequent Ar+ low energy ion milling at 500 eV. (b) Atomic resolution image taken from the area indicated in a, showing ~2-3 nm large SFTs imaged along the (110) projection.

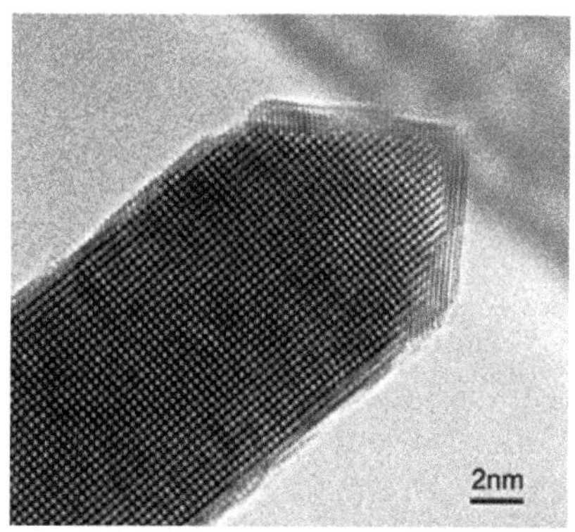

High resolution TEM image of RuO2 nanorod.

TEM: High Resolution Cross View

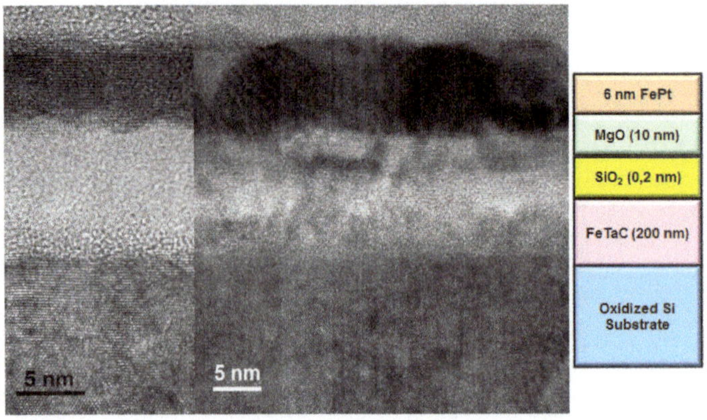

Cross-sectional TEM image of FePt based double-layered perpendicular recording media.

Figure shows the role of thin SiO2 layer on top of FeTaC soft layer to control the growth of the MgO into a particular direction. Left image shows the growth of MgO as amorphous like structure and right images shows the MgO(200) structure after the introduction of 2 nm SiO2 on top of FeTaC layer.

Energy Dispersive Spectroscopy (or) Energy Dispersive X-ray Analysis

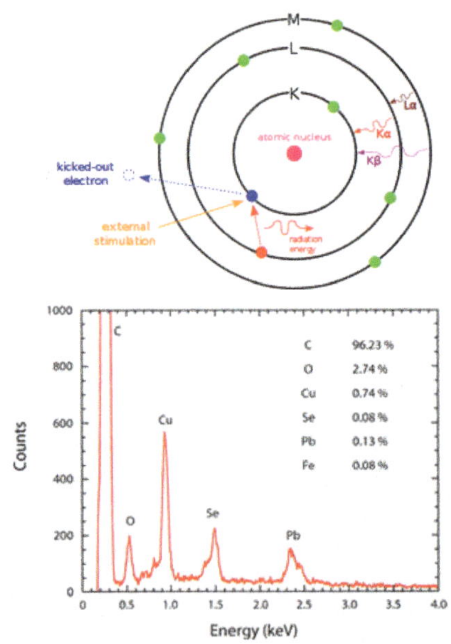

Schematic of X-ray generation and the plot of intensity versus X-ray energy to quantify the content. EDS data for the PbSe -MWNT sample. Copper and iron are from the formvar TEM grid and residual nanotube growth catalyst respectively.

- When the electron interacts with the specimen, the X-ray characteristics of atoms in the irradiated area are emitted, as shown in figure.
- By analyzing the binding energies, the nature of atom or element can be identified.
- By counting the numbers of X-rays emitted, the concentration of atoms in the spectrum can be determined.

- This method helps in analyzing the composition of the specimen.

TEM: Electron Energy Loss Spectroscopy (EELS)

- In electron energy loss spectroscopy (EELS), developed by James Hillier and RF Baker in mid 1940s, a material is exposed to a beam of electrons with a known and narrow range of kinetic energies, as shown in figure. As a result, some of the electrons undergo inelastic scattering, i.e., they lose energy and have their paths deflected randomly. The amount of energy loss can be measured via an electron spectrometer and interpreted in terms of what caused the energy loss.

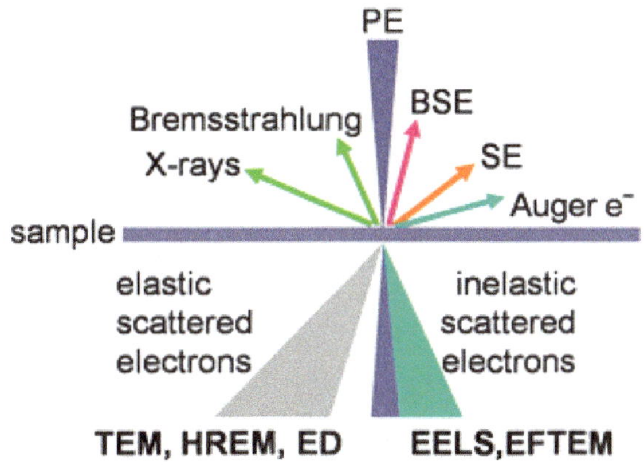

Schematic of electron interaction with atoms in materials and the consequent process.

- In addition, inelastic interactions include phonon excitations, inter/intra band transitions, inner shell ionizations, plasmon excitations, etc. The inner-shell ionizations are particularly useful for detecting the elemental compositions of a material.

- For example, one might observe that a larger-than-expected number of electrons comes through the material with energy of 285 eV less than they had when they entered the material.

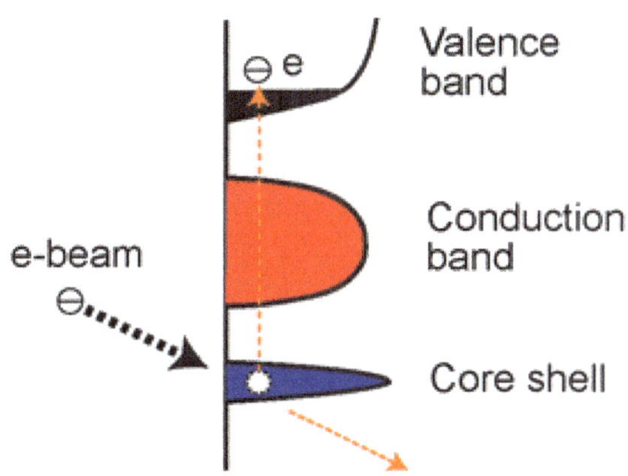

Schematic of energy loss induced by electron transition.

- It so happens that this is about the amount of energy needed to remove an inner-shell electron from a carbon atom. This can be taken as evidence that there is a significant amount of carbon in the part of the material that is being hit by the electron beam.

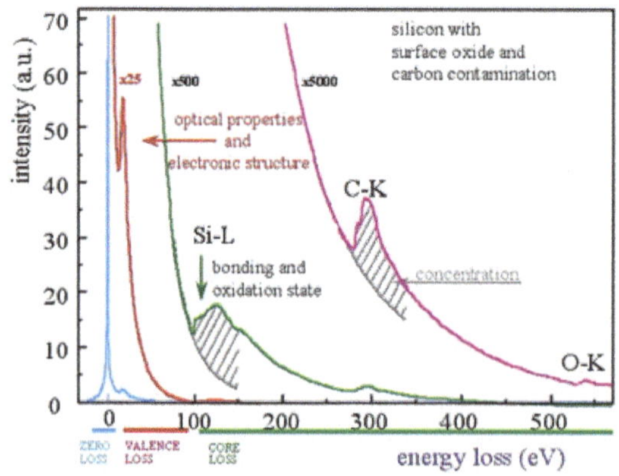

The plot of intensity versus energy loss for various materials.

- With some care, and looking at a wide range of energy losses, one can determine the types of atoms, and the numbers of atoms of each type, being struck by the beam. The scattering angle can also be measured, providing information about dispersion relation of whatever material excitation caused the inelastic scattering.

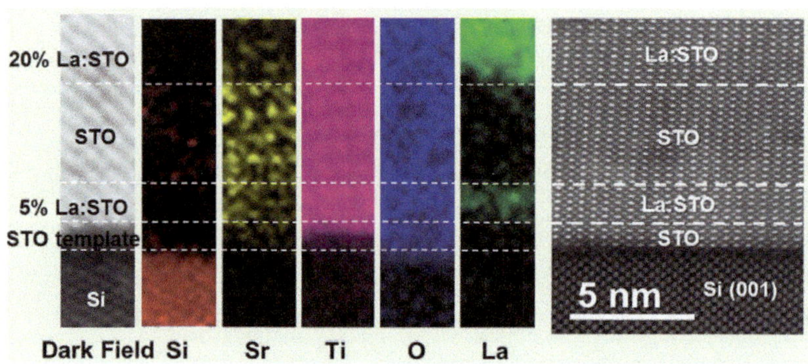

EELS images and cross-sectional TEM images of multilayer structure.

Figure displays the compositional mapping using electron energy-loss spectroscopy and highlighting the control of La distribution, and (right) high-resolution cross-sectional TEM image showing the clean interface between Si-STO as well as STO and the La:STO layers. The existence of constituent materials in each layer without interlayer mixing is also apparent.

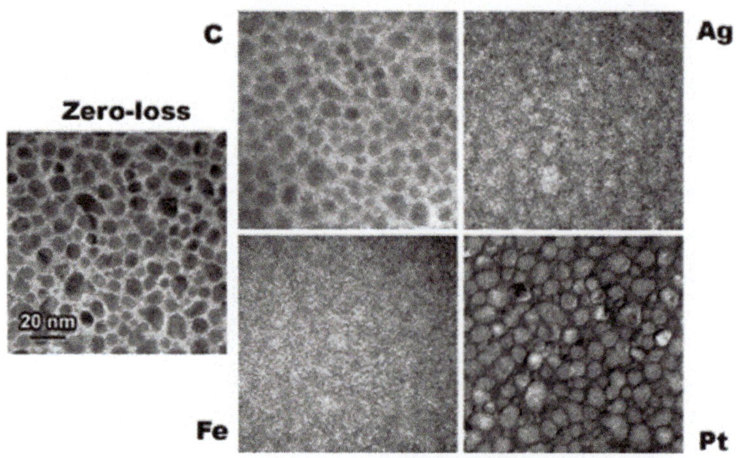

Compositional mapping of $(FePt)_{0.8}Ag_{0.2}$-50vol%C film.

Figure displays the energy filtered TEM images of the (FePt)0.8Ag0.2-50vol%C film. It is clearly identifiable that the C is majorly at the grain boundary, while the Ag, Fe,and Pt are existing majorly within the grain.

There are much more options such as Lorentz domain mapping, high-angle annular dark field, scanning tunneling electron microscope, etc available in modern microscopes. Interested readers may go through additional materials for more information.

TEM: (Drawbacks)

- Expertise: The interpretation of the images and spectrometry data requires expertise. Typically, a combination of education in physics and experience in interpreting TEM data is very much necessary.
- Expense of the Instrument: TEM is relatively expensive instruments.
- Sample Preparation: TEM sample preparation is still very much an art. The ability to polish a specimen to electron transparency evenly over several tens of microns takes a good deal of practice and patience.

Scanning Electron Microscope

A scanning electron microscope (SEM) is a type of electron microscope that produces images of a sample by scanning it with a focused beam of electrons. The electrons interact with atoms in the sample, producing various signals that contain information about the sample's surface topography and composition. The electron beam is generally scanned in a raster scan pattern, and the beam's position is combined with the detected signal to produce an image. SEM can achieve resolution better than 1 nanometer. Specimens can be observed in high vacuum, in low vacuum, in wet conditions (in environmental SEM), and at a wide range of cryogenic or elevated temperatures.

These pollen grains taken on an SEM show the characteristic depth of field of SEM micrographs.

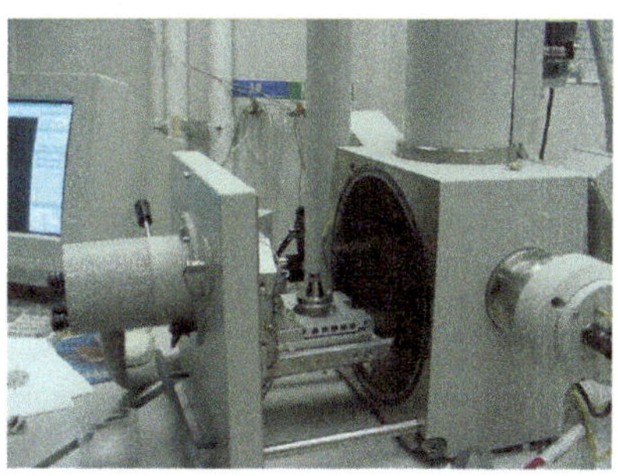

SEM opened sample chamber

Analog type SEM

The most common SEM mode is detection of secondary electrons emitted by atoms excited by the electron beam. The number of secondary electrons that can be detected depends, among other things, on specimen topography. By scanning the sample and collecting the secondary electrons that are emitted using a special detector, an image displaying the topography of the surface is created.

History

An account of the early history of SEM has been presented by McMullan. Although Max Knoll produced a photo with a 50 mm object-field-width showing channeling contrast by the use of an electron beam scanner, it was Manfred von Ardenne who in 1937 invented a true microscope with high magnification by scanning a very small raster with a demagnified and finely focused electron beam. Ardenne applied the scanning principle not only to achieve magnification but also to purposefully eliminate the chromatic aberration otherwise inherent in the electron microscope. He further discussed the various detection modes, possibilities and theory of SEM, together with the construction of the first high magnification SEM. Further work was reported by Zworykin's group, followed by the Cambridge groups in the 1950s and early 1960s headed by Charles Oatley, all of which finally led to the marketing of the first commercial instrument by Cambridge Scientific Instrument Company as the "Stereoscan" in 1965, which was delivered to DuPont.

Principles and Capacities

The types of signals produced by an SEM include secondary electrons (SE), reflected or back-scattered electrons (BSE), photons of characteristic X-rays and light (cathodoluminescence) (CL), absorbed current (specimen current) and transmitted electrons. Secondary electron detectors are standard equipment in all SEMs, but it is rare that a single machine would have detectors for all other possible signals.

The signals result from interactions of the electron beam with atoms at various depths within the sample. In the most common or standard detection mode, ie secondary electron imaging or SEI, the secondary electrons are emitted from very close to the specimen surface. Consequently, SEM can produce very high-resolution images of a sample surface, revealing details less than 1 nm in size. Back-scattered electrons (BSE) are beam electrons that are reflected from the sample by elastic scattering. They emerge from deeper locations within the specimen and consequently the resolution of BSE images is generally poorer than SE images. However, BSE are often used in analytical SEM along with the spectra made from the characteristic X-rays, because the intensity of the BSE signal is strongly related to the atomic number (Z) of the specimen. BSE images can provide information about the distribution of different elements in the sample. For the same reason, BSE imaging can image colloidal gold immuno-labels of 5 or 10 nm diameter, which would otherwise be difficult or impossible to detect in secondary electron images in biological specimens. Characteristic X-rays are emitted when the electron beam removes an inner shell electron from the sample, causing a higher-energy electron to fill the shell and release energy. These characteristic X-rays are used to identify the composition and measure the abundance of elements in the sample.

Due to the very narrow electron beam, SEM micrographs have a large depth of field yielding a characteristic three-dimensional appearance useful for understanding the surface structure of a sample. A wide range of magnifications is possible, from about 10 times (about equivalent to that of a powerful hand-lens) to more than 500,000 times, about 250 times the magnification limit of the best light microscopes.

Sample Preparation

Samples for SEM can be solid, bulk specimens of any size that will fit within the specimen chamber. Samples are generally mounted rigidly on a specimen holder called a specimen stub using a conductive adhesive, but mechanical clamping is a possible alternative. Several models of SEM can examine any part of a 6-inch (15 cm) semiconductor wafer, and some can tilt an object of that size to 45° and provide continuous 360° rotation.

A spider sputter-coated in gold, having been prepared for viewing with an SEM.

Nonconductive specimens tend to charge when scanned by the electron beam, and especially in secondary electron imaging mode, this causes scanning faults and other image artifacts. For conventional imaging in the SEM, specimens must be electrically conductive, at least at the surface, and electrically grounded to prevent the accumulation of electrostatic charge. Metal objects require little special preparation for SEM except for cleaning and mounting on a specimen stub. Non-conducting materials are usually coated with an ultrathin coating of electrically conducting material, deposited on the sample either by low-vacuum sputter coating or by high-vacuum evaporation. Conductive materials in current use for specimen coating include gold, gold/palladium alloy, platinum, osmium, iridium, tungsten, chromium, and graphite. Coating with heavy metals may increase signal/noise ratio for samples of low atomic number (Z). The improvement arises because secondary electron emission for high-Z materials is enhanced.

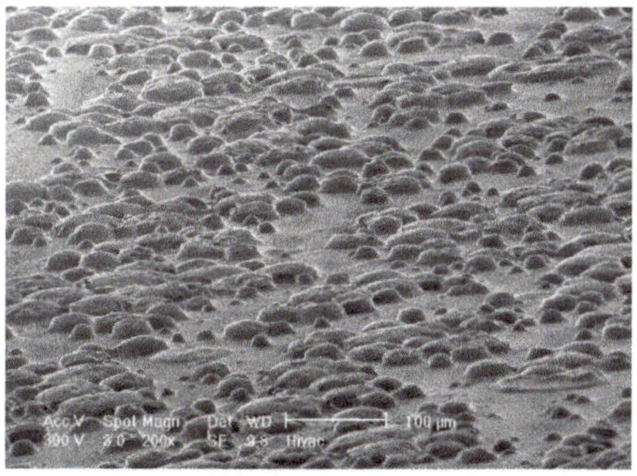

Low-voltage micrograph (300 V) of distribution of adhesive droplets on a Post-it note.
No conductive coating was applied: such a coating would alter this fragile specimen.

An alternative to coating for some biological samples is to increase the bulk conductivity of the material by impregnation with osmium using variants of the OTO staining method (O-osmium tetroxide, T-thiocarbohydrazide, O-osmium).

Nonconducting specimens may be imaged without coating using an environmental SEM (ESEM) or low-voltage mode of SEM operation. In ESEM instruments the specimen is placed in a relatively high-pressure chamber and the electron optical column is differentially pumped to keep vacuum adequately low at the electron gun. The high-pressure region around the sample in the ESEM neutralizes charge and provides an amplification of the secondary electron signal. Low-voltage SEM is typically conducted in an FEG-SEM because field emission guns (FEG) are capable of producing high primary electron brightness and small spot size even at low accelerating potentials. To prevent charging of non-conductive specimens, operating conditions must be adjusted such that the incoming beam current is equal to sum of outcoming secondary and backscattered electrons currents a condition that is more often met at accelerating voltages of 0.3–4 kV.

Synthetic replicas can be made to avoid the use of original samples when they are not suitable or available for SEM examination due to methodological obstacles or legal issues. This technique is achieved in two steps: (1) a mold of the original surface is made using a silicone-based dental elastomer, and (2) a replica of the original surface is obtained by pouring a synthetic resin into the mold.

Embedding in a resin with further polishing to a mirror-like finish can be used for both biological and materials specimens when imaging in backscattered electrons or when doing quantitative X-ray microanalysis.

The main preparation techniques are not required in the environmental SEM outlined below, but some biological specimens can benefit from fixation.

Biological Samples

For SEM, a specimen is normally required to be completely dry, since the specimen chamber is at high vacuum. Hard, dry materials such as wood, bone, feathers, dried insects, or shells (including egg shells) can be examined with little further treatment, but living cells and tissues and whole, soft-bodied organisms usually require chemical fixation to preserve and stabilize their structure.

Fixation is usually performed by incubation in a solution of a buffered chemical fixative, such as glutaraldehyde, sometimes in combination with formaldehyde and other fixatives, and optionally followed by postfixation with osmium tetroxide. The fixed tissue is then dehydrated. Because air-drying causes collapse and shrinkage, this is commonly achieved by replacement of water in the cells with organic solvents such as ethanol or acetone, and replacement of these solvents in turn with a transitional fluid such as liquid carbon dioxide by critical point drying. The carbon dioxide is finally removed while in a supercritical state, so that no gas–liquid interface is present within the sample during drying.

The dry specimen is usually mounted on a specimen stub using an adhesive such as epoxy resin or electrically conductive double-sided adhesive tape, and sputter-coated with gold or gold/palladium alloy before examination in the microscope. Other samples may require sectioning (with a microtome) and staining.

If the SEM is equipped with a cold stage for cryo microscopy, cryofixation may be used and low-temperature scanning electron microscopy performed on the cryogenically fixed specimens. Cryo-fixed specimens may be cryo-fractured under vacuum in a special apparatus to reveal internal structure, sputter-coated, and transferred onto the SEM cryo-stage while still frozen. Low-temperature scanning electron microscopy (LT-SEM) is also applicable to the imaging of temperature-sensitive materials such as ice and fats.

Freeze-fracturing, freeze-etch or freeze-and-break is a preparation method particularly useful for examining lipid membranes and their incorporated proteins in "face on" view. The preparation method reveals the proteins embedded in the lipid bilayer.

Materials

Back-scattered electron imaging, quantitative X-ray analysis, and X-ray mapping of specimens often requires grinding and polishing the surfaces to an ultra smooth surface. Specimens that undergo WDS or EDS analysis are often carbon-coated. In general, metals are not coated prior to imaging in the SEM because they are conductive and provide their own pathway to ground.

Fractography is the study of fractured surfaces that can be done on a light microscope or, commonly, on an SEM. The fractured surface is cut to a suitable size, cleaned of any organic residues, and mounted on a specimen holder for viewing in the SEM.

Integrated circuits may be cut with a focused ion beam (FIB) or other ion beam milling instrument for viewing in the SEM. The SEM in the first case may be incorporated into the FIB.

Metals, geological specimens, and integrated circuits all may also be chemically polished for viewing in the SEM.

Special high-resolution coating techniques are required for high-magnification imaging of inorganic thin films.

Scanning Process and Image Formation

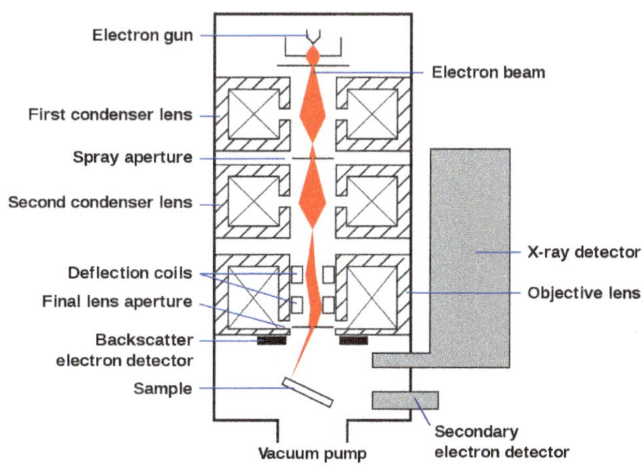

Schematic of an SEM.

In a typical SEM, an electron beam is thermionically emitted from an electron gun fitted with a tungsten filament cathode. Tungsten is normally used in thermionic electron guns because it has the highest melting point and lowest vapor pressure of all metals, thereby allowing it to be electrically heated for electron emission, and because of its low cost. Other types of electron emitters include lanthanum hexaboride (LaB_6) cathodes, which can be used in a standard tungsten filament SEM if the vacuum system is upgraded or field emission guns (FEG), which may be of the cold-cathode type using tungsten single crystal emitters or the thermally assisted Schottky type, that use emitters of zirconium oxide.

The electron beam, which typically has an energy ranging from 0.2 keV to 40 keV, is focused by one or two condenser lenses to a spot about 0.4 nm to 5 nm in diameter. The beam passes through pairs of scanning coils or pairs of deflector plates in the electron column, typically in the final lens, which deflect the beam in the x and y axes so that it scans in a raster fashion over a rectangular area of the sample surface.

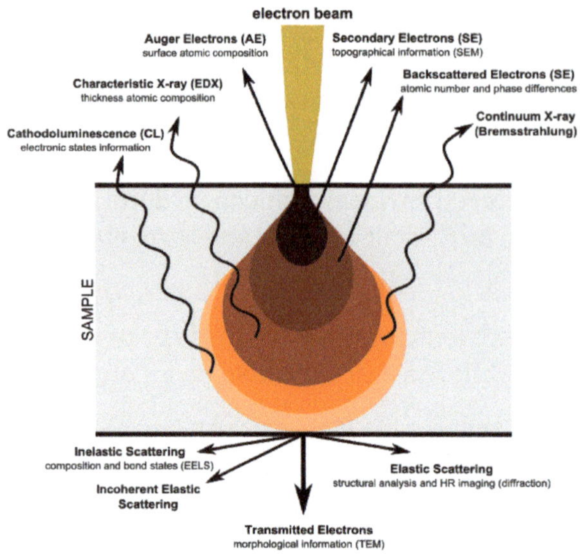

Signals emitted from different parts of the interaction volume

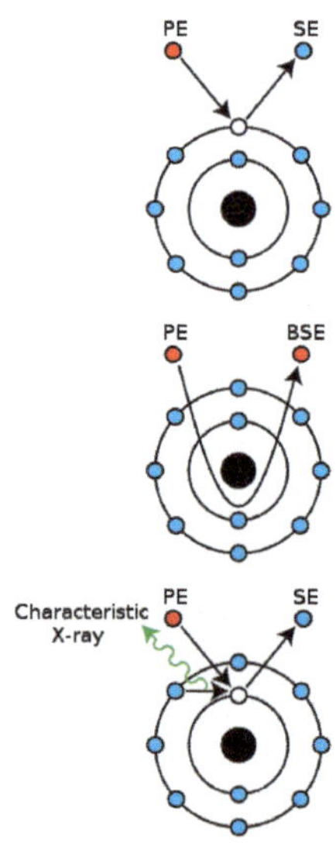

Mechanisms of emission of secondary electrons, backscattered electrons, and characteristic X-rays from atoms of the sample

When the primary electron beam interacts with the sample, the electrons lose energy by repeated random scattering and absorption within a teardrop-shaped volume of the specimen known as the interaction volume, which extends from less than 100 nm to approximately 5 µm into the surface. The size of the interaction volume depends on the electron's landing energy, the atomic number of the specimen and the specimen's density. The energy exchange between the electron beam and the sample results in the reflection of high-energy electrons by elastic scattering, emission of secondary electrons by inelastic scattering and the emission of electromagnetic radiation, each of which can be detected by specialized detectors. The beam current absorbed by the specimen can also be detected and used to create images of the distribution of specimen current. Electronic amplifiers of various types are used to amplify the signals, which are displayed as variations in brightness on a computer monitor (or, for vintage models, on a cathode ray tube). Each pixel of computer video memory is synchronized with the position of the beam on the specimen in the microscope, and the resulting image is therefore a distribution map of the intensity of the signal being emitted from the scanned area of the specimen. In older microscopes images may be captured by photography from a high-resolution cathode ray tube, but in modern machines they are digitised and saved as digital images.

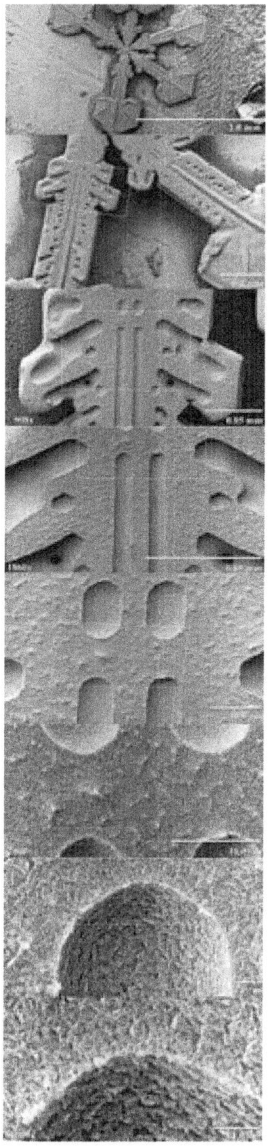

Low-temperature SEM magnification series for a snow crystal. The crystals are captured, stored, and sputter-coated with platinum at cryogenic temperatures for imaging.

Magnification

Magnification in an SEM can be controlled over a range of about 6 orders of magnitude from about 10 to 500,000 times. Unlike optical and transmission electron microscopes, image magnification in an SEM is not a function of the power of the objective lens. SEMs may have condenser and objective lenses, but their function is to focus the beam to a spot, and not to image the specimen. Provided the electron gun can generate a beam with sufficiently small diameter, an SEM could in principle work entirely without condenser or objective lenses, although it might not be very versatile or achieve very high resolution. In an SEM, as in scanning probe microscopy, magnification results from the ratio of the dimensions of the raster on the specimen and the raster on the display device. Assuming that the display screen has a fixed size, higher magnification results from reducing the size of the raster on the specimen, and vice versa. Magnification is therefore controlled by the current supplied to the x, y scanning coils, or the voltage supplied to the x, y deflector plates, and not by objective lens power.

Detection of Secondary Electrons

The most common imaging mode collects low-energy (<50 eV) secondary electrons that are ejected from the k-shell of the specimen atoms by inelastic scattering interactions with beam electrons. Due to their low energy, these electrons originate within a few nanometers from the sample surface. The electrons are detected by an Everhart-Thornley detector, which is a type of scintillator-photomultiplier system. The secondary electrons are first collected by attracting them towards an electrically biased grid at about +400 V, and then further accelerated towards a phosphor or scintillator positively biased to about +2,000 V. The accelerated secondary electrons are now sufficiently energetic to cause the scintillator to emit flashes of light (cathodoluminescence), which are conducted to a photomultiplier outside the SEM column via a light pipe and a window in the wall of the specimen chamber. The amplified electrical signal output by the photomultiplier is displayed as a two-dimensional intensity distribution that can be viewed and photographed on an analogue video display, or subjected to analog-to-digital conversion and displayed and saved as a digital image. This process relies on a raster-scanned primary beam. The brightness of the signal depends on the number of secondary electrons reaching the detector. If the beam enters the sample perpendicular to the surface, then the activated region is uniform about the axis of the beam and a certain number of electrons "escape" from within the sample. As the angle of incidence increases, the interaction volume increases and the "escape" distance of one side of the beam decreases, resulting in more secondary electrons being emitted from the sample. Thus steep surfaces and edges tend to be brighter than flat surfaces, which results in images with a well-defined, three-dimensional appearance. Using the signal of secondary electrons image resolution less than 0.5 nm is possible.

Detection of Backscattered Electrons

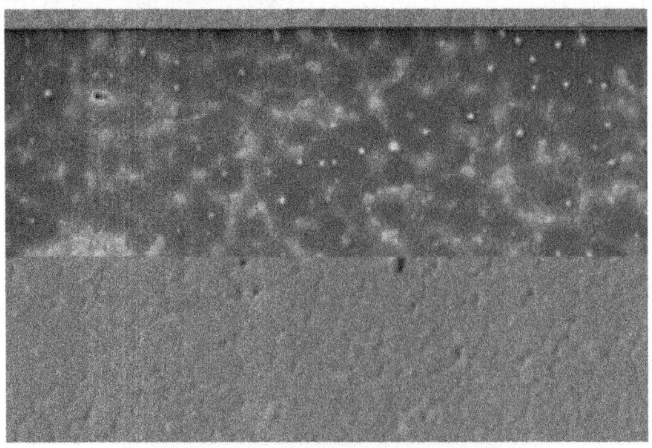

Comparison of SEM techniques:
Top: backscattered electron analysis – composition
Bottom: secondary electron analysis – topography

Backscattered electrons (BSE) consist of high-energy electrons originating in the electron beam, that are reflected or back-scattered out of the specimen interaction volume by elastic scattering interactions with specimen atoms. Since heavy elements (high atomic number) backscatter electrons more strongly than light elements (low atomic number), and thus appear brighter in the image, BSE are used to detect contrast between areas with different chemical compositions. The Everhart-Thornley detector, which is normally positioned to one side of the specimen, is inefficient for the detection of backscattered electrons because few such electrons are emitted in the solid angle subtended by the detector, and because the positively biased detection grid has little ability to attract the higher energy BSE. Dedicated backscattered electron detectors are positioned above the sample in a "doughnut" type arrangement, concentric with the electron beam, maximizing the solid angle of collection. BSE detectors are usually either of scintillator or of semiconductor types. When all parts of the detector are used to collect electrons symmetrically about the beam, atomic number contrast is produced. However, strong topographic contrast is produced by collecting back-scattered electrons from one side above the specimen using an asymmetrical, directional BSE detector; the resulting contrast appears as illumination of the topography from that side. Semiconductor detectors can be made in radial segments that can be switched in or out to control the type of contrast produced and its directionality.

Backscattered electrons can also be used to form an electron backscatter diffraction (EBSD) image that can be used to determine the crystallographic structure of the specimen.

Beam-injection Analysis of Semiconductors

The nature of the SEM's probe, energetic electrons, makes it uniquely suited to examining the optical and electronic properties of semiconductor materials. The high-energy electrons from the SEM beam will inject charge carriers into the semiconductor. Thus, beam electrons lose energy by promoting electrons from the valence band into the conduction band, leaving behind holes.

In a direct bandgap material, recombination of these electron-hole pairs will result in cathodoluminescence; if the sample contains an internal electric field, such as is present at a p-n junction, the SEM beam injection of carriers will cause electron beam induced current (EBIC) to flow. Cathodoluminescence and EBIC are referred to as "beam-injection" techniques, and are very powerful probes of the optoelectronic behavior of semiconductors, in particular for studying nanoscale features and defects.

Cathodoluminescence

Cathodoluminescence, the emission of light when atoms excited by high-energy electrons return to their ground state, is analogous to UV-induced fluorescence, and some materials such as zinc sulfide and some fluorescent dyes, exhibit both phenomena. Over the last decades, cathodoluminescence was most commonly experienced as the light emission from the inner surface of the cathode ray tube in television sets and computer CRT monitors. In the SEM, CL detectors either collect all light emitted by the specimen or can analyse the wavelengths emitted by the specimen and display an emission spectrum or an image of the distribution of cathodoluminescence emitted by the specimen in real color.

X-ray Microanalysis

Characteristic X-rays that are produced by the interaction of electrons with the sample may also be detected in an SEM equipped for energy-dispersive X-ray spectroscopy or wavelength dispersive X-ray spectroscopy. Analysis of the x-ray signals may be used to map the distribution and estimate the abundance of elements in the sample.

Resolution of the SEM

SEM is not a camera and the detector is not continuously image-forming like a CCD array or film. Unlike in an optical system, the resolution is not limited by the diffraction limit, fineness of lenses or mirrors or detector array resolution. The focusing optics can be large and coarse, and the SE detector is fist-sized and simply detects current. Instead, the spatial resolution of the SEM depends on the size of the electron spot, which in turn depends on both the wavelength of the electrons and the electron-optical system that produces the scanning beam. The resolution is also limited by the size of the interaction volume, the volume of specimen material that interacts with the electron beam. The spot size and the interaction volume are both large compared to the distances between atoms, so the resolution of the SEM is not high enough to image individual atoms, as is possible with transmission electron microscope (TEM). The SEM has compensating advantages, though, including the ability to image a comparatively large area of the specimen; the ability to image bulk materials (not just thin films or foils); and the variety of analytical modes available for measuring the composition and properties of the specimen. Depending on the instrument, the resolution can fall somewhere between less than 1 nm and 20 nm. As of 2009, The world's highest resolution conventional (<30 kV) SEM can reach a point resolution of 0.4 nm using a secondary electron detector.

Environmental SEM

Conventional SEM requires samples to be imaged under vacuum, because a gas atmosphere rapidly spreads and attenuates electron beams. As a consequence, samples that produce a significant amount of vapour, e.g. wet biological samples or oil-bearing rock, must be either dried or cryogenically frozen. Processes involving phase transitions, such as the drying of adhesives or melting of alloys, liquid transport, chemical reactions, and solid-air-gas systems, in general cannot be observed. Some observations of living insects have been possible however.

The first commercial development of the ESEM in the late 1980s allowed samples to be observed in low-pressure gaseous environments (e.g. 1–50 Torr or 0.1–6.7 kPa) and high relative humidity (up to 100%). This was made possible by the development of a secondary-electron detector capable of operating in the presence of water vapour and by the use of pressure-limiting apertures with differential pumping in the path of the electron beam to separate the vacuum region (around the gun and lenses) from the sample chamber.

The first commercial ESEMs were produced by the ElectroScan Corporation in USA in 1988. ElectroScan was taken over by Philips (who later sold their electron-optics division to FEI Company) in 1996.

ESEM is especially useful for non-metallic and biological materials because coating with carbon or gold is unnecessary. Uncoated Plastics and Elastomers can be routinely examined, as can uncoated biological samples. Coating can be difficult to reverse, may conceal small features on the surface of the sample and may reduce the value of the results obtained. X-ray analysis is difficult with a coating of a heavy metal, so carbon coatings are routinely used in conventional SEMs, but ESEM makes it possible to perform X-ray microanalysis on uncoated non-conductive specimens; however some specific for ESEM artifacts are introduced in X-ray analysis. ESEM may be the preferred for electron microscopy of unique samples from criminal or civil actions, where forensic analysis may need to be repeated by several different experts.

Transmission SEM

The SEM can also be used in transmission mode by simply incorporating an appropriate detector below a thin specimen section. Both bright and dark field imaging has been reported in the generally low accelerating beam voltage range used in SEM, which increases the contrast of unstained biological specimens at high magnifications with a field emission electron gun. This mode of operation has been abbreviated by the acronym tSEM.

Color in SEM

Electron microscopes do not naturally produce color images, as a SEM produces a single value per pixel; this value corresponds to the number of electrons received by the detector during a small period of time of the scanning when the beam is targeted to the (x,y) pixel position.

This single number is usually represented, for each pixel, by a grey level, forming a "black-and-white" image. However, several ways have been used to get color electron microscopy images.

False Color Using a Single Detector

- On compositional images of flat surfaces (typically BSE):

The easiest way to get color is to associate to this single number an arbitrary color, using a color look-up table (i.e. each grey level is replaced by a chosen color). This method is known as false color. On a BSE image, false color may be performed to better distinguish the various phases of the sample.

- On textured-surface images:

As an alternative to simply replacing each grey level by a color, a sample observed by an oblique beam allows to create an approximative topography image. Such topography can then be processed by 3D-rendering algorithms for a more natural rendering of the surface texture.

Surface of a kidney stone

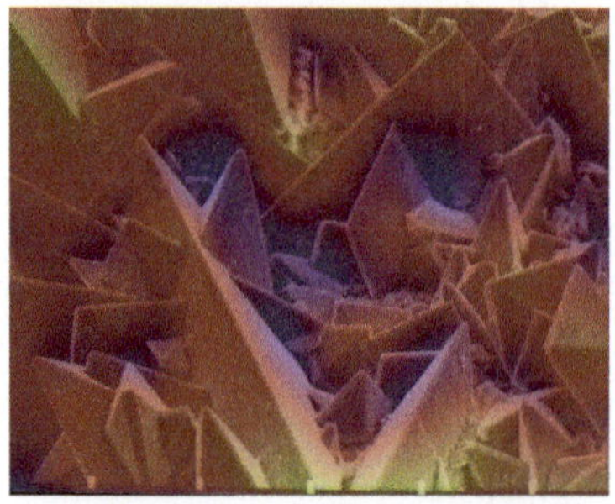

The same after re-processing of the color from the estimated topography

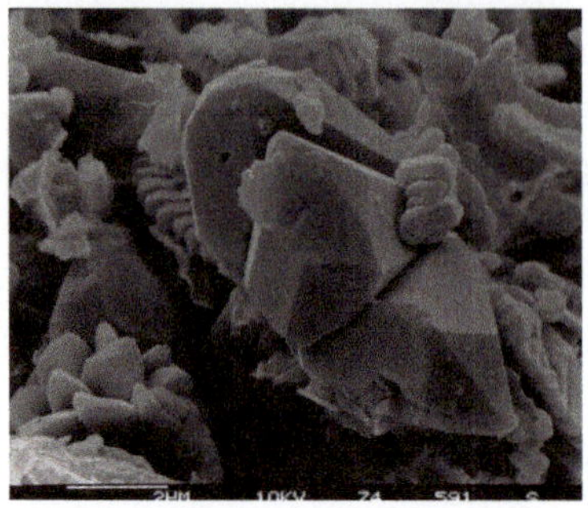

SEM image of a diagenetically altered discoaster

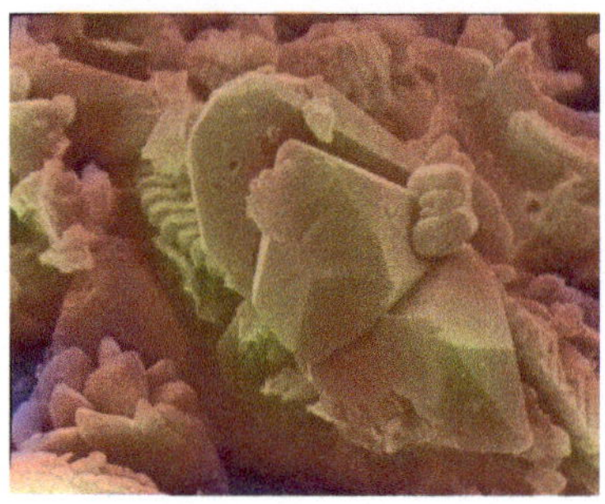

The same image after similar colorization

SEM Image Coloring

Very often, published SEM images are artificially colored. This may be done for aesthetic effect, to clarify structure or to add a realistic appearance to the sample and generally does not add information about the specimen.

Coloring may be performed manually with photo-editing software, or semi-automatically with dedicated software using feature-detection or object-oriented segmentation.

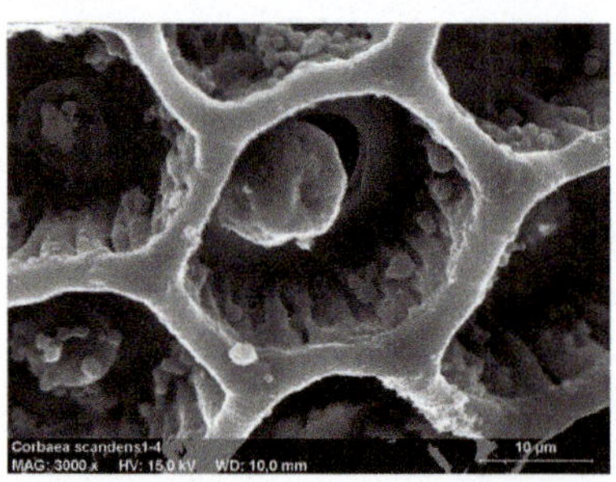

SEM image of Cobaea scandens pollen

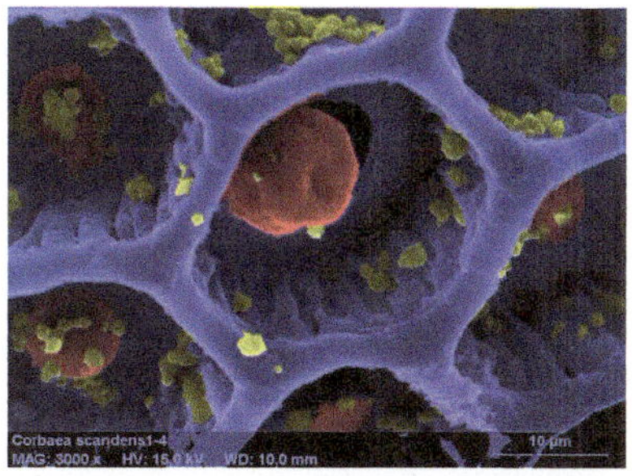

The same after semi-automatic coloring. Arbitrary colors help identifying the various elements of the structure

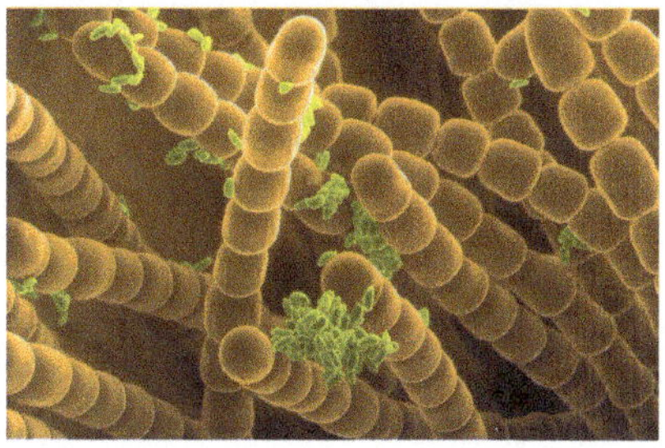

Colored SEM image of *Tradescantia* pollen and stamens

Color Built Using Multiple Electron Detectors

In some configurations more information is gathered per pixel, often by the use of multiple detectors.

As a common example, secondary electron and backscattered electron detectors are superimposed and a color is assigned to each of the images captured by each detector, with an end result of a combined color image where colors are related to the density of the components. This method is known as density-dependent color SEM (DDC-SEM). Micrographs produced by DDC-SEM retain topographical information, which is better captured by the secondary electrons detector and combine it to the information about density, obtained by the backscattered electron detector.

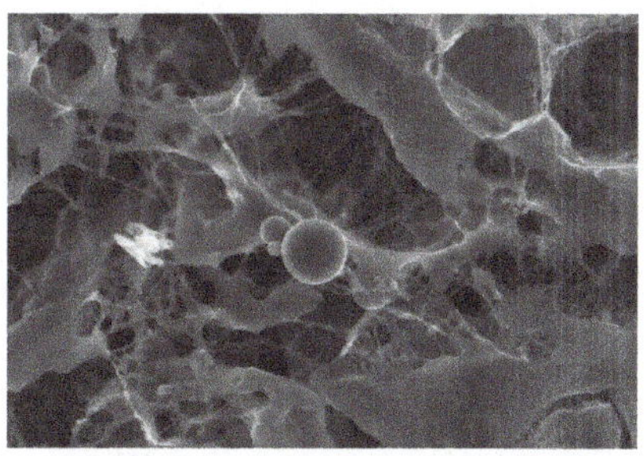

DDC-SEM of calcified particle in cardiac tissue - Signal 1 : SE

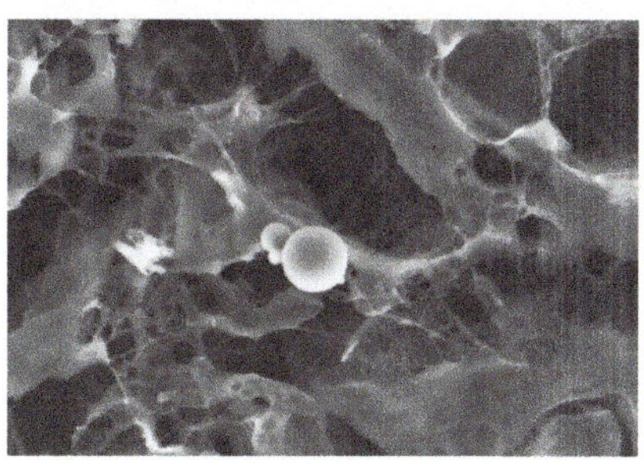

Signal 2 : BSE

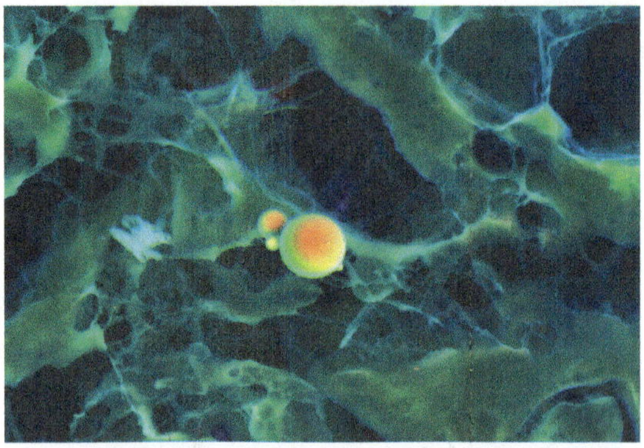

Colorized image obtained from the two previous. Density-dependent color scanning electron micrograph SEM (DDC-SEM) of cardiovascular calcification, showing in orange a calcium phosphate spherical particle (denser material) and, in green, the extracellular matrix (less dense material)

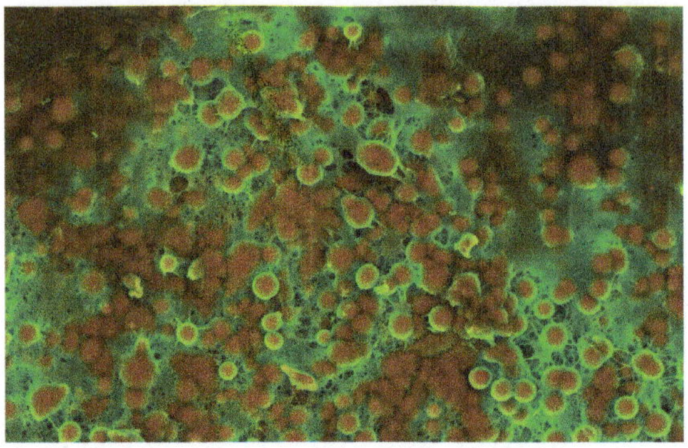

Same work with a larger view, part of a study on human cardiovascular tissue calcification

Analytical Signals Based on Generated Photons

Measurement of the energy of photons emitted from the specimen is a common method to get analytical capabilities. Examples are the Energy-dispersive X-ray spectroscopy (EDS) detectors used in elemental analysis and Cathodoluminescence microscope (CL) systems that analyse the intensity and spectrum of electron-induced luminescence in (for example) geological specimens. In SEM systems using these detectors it is common to color code these extra signals and superimpose them in a single color image, so that differences in the distribution of the various components of the specimen can be seen clearly and compared. Optionally, the standard secondary electron image can be merged with the one or more compositional channels, so that the specimen's structure and composition can be compared. Such images can be made while maintaining the full integrity of the original signal data, which is not modified in any way.

3D in SEM

SEMs do not naturally provide 3D images contrary to SPMs. However 3D data can be obtained using an SEM with different methods as follows.

3D SEM Reconstruction from a Stereo Pair

- photogrammetry (2 or 3 images from tilted specimen)

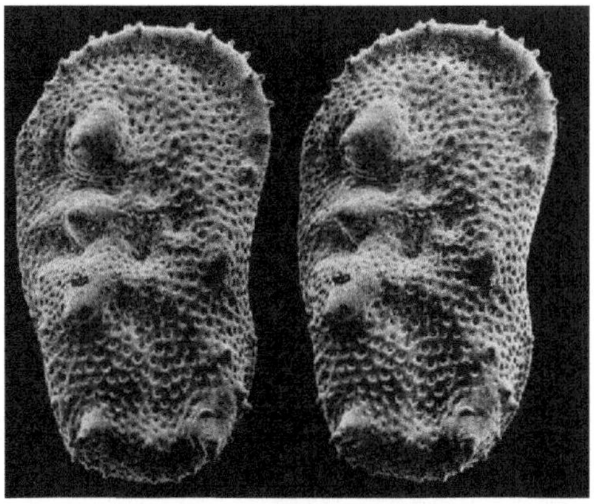

An SEM stereo pair of microfossils of less than 1 mm in size (Ostracoda) produced by tilting along the longitudinal axis.

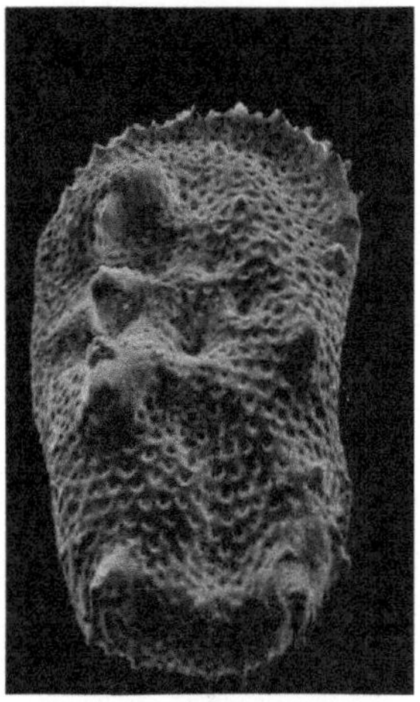

From this pair of SEM images, the third dimension has been reconstructed by photogrammetry (using MountainsMap software) ; then a series of 3D representations with different angles have been made and assembled into a GIF file to produce this animation.

Photometric 3D SEM Reconstruction from a Four-quadrant Detector "Shape from Shading"

This method typically uses a four-quadrant BSE detector. The microscope produces four images of the same specimen at the same time, so no tilt is required. The method gives metrological 3D dimensions as far as the slope of the specimen remains reasonable. As it works by integration of the slope, vertical slopes and overhangs are ignored ; for instance, if an entire sphere lies on a flat, only the main part of the upper hemisphere is seen emerging above the flat, resulting in wrong altitude of the sphere apex.

Photometric 3D Rendering from a Single SEM Image

This method requires an SEM image obtained in oblique low angle lighting. The grey-level is then interpreted as the slope, and the slope integrated to restore the specimen topography. This method is interesting for visual enhancement and the detection of the shape and position of objects ; however the vertical heights cannot usually be calibrated, contrary to other methods such as photogrammetry.

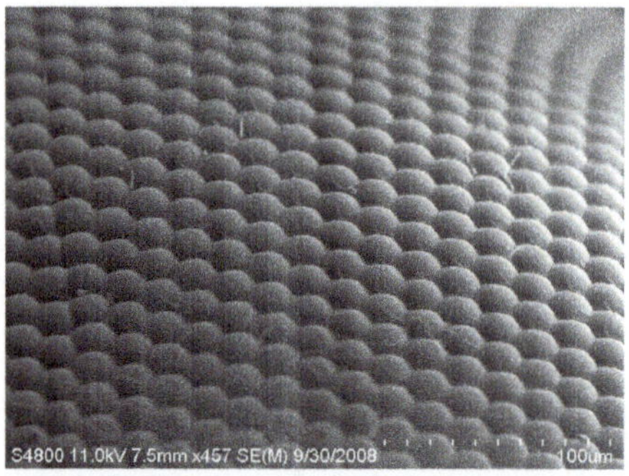

SEM image of a house fly compound eye surface at 450× magnification.

Detail of the previous image.

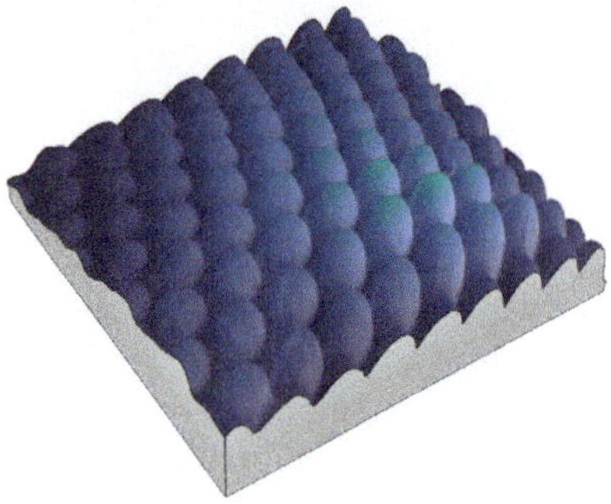

SEM 3D reconstruction from the previous using shape from shading algorithms.

Same as the previous, but with lighting homogenized before applying the shape from shading algorithms

Other Types of 3D SEM Reconstruction

- inverse reconstruction using electron-material interactive models
- vertical stacks of SEM micrographs plus image-processing software

- Multi-Resolution reconstruction using single 2D File: High-quality 3D imaging may be an ultimate solution for revealing the complexities of any porous media, but acquiring them is costly and time consuming. High-quality 2D SEM images, on the other hand, are widely available. Recently, a novel three-step, multiscale, multiresolution reconstruction method is presented that directly uses 2D images in order to develop 3D models. This method, based on a Shannon Entropy and conditional simulation, can be used for most of the available stationary materials and can build various stochastic 3D models just using a few thin sections.

Applications of 3D SEM

Possible applications are roughness measurement, measurement of fractal dimension, examining fracture surface of metals, characterization of materials, corrosion measurement, and dimensional measurements at the nano scale (step height, volume, angle, flatness, bearing ratio, coplanarity, etc.).

SEM: Principles

- When the electron beam hits the sample, the interaction of the beam electrons from the filament and the sample atoms generates a variety of signals.
- Depending on the sample, these can include secondary electrons (electrons from the sample itself), backscattered electrons (beam electrons from the filament that bounce off nuclei of atoms in the sample), X-rays, light, heat, and even transmitted electrons (beam electrons that pass through the sample).

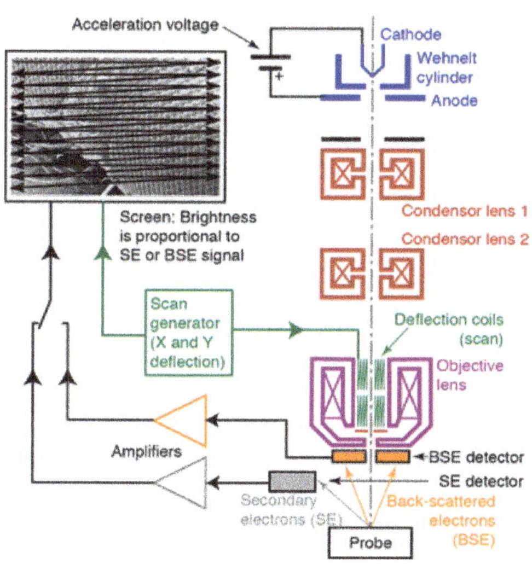

Schematic of SEM set up.

- Figure shows the schematic representation of the SEM. The principle elements are electron gun, condenser lens, scanning or deflection coils.

SEM has an electron column as shown in figure, where the electrons are generated and arranged through electric and magnetic fields in a proper direction to achieve the required incident beam at the sample surface.

- The whole process is operated under vacuum to avoid collision between electrons and gas molecules.

Photoview of interior of SEM.

When the electron beam hits the sample, the interaction of the beam electrons from the filament and the sample atoms generates a variety of signals.

Hence, the SEM has several detectors to view the electron signals from the sample. Two of these detectors can be seen from inside the specimen chamber. The secondary electron detector used for detecting the secondary electrons looks like a Faraday cage (1). The backscattered electron detector is located above the sample (2) to detect backscattered electrons.

- Unlike the light microscope (in which light forms an instant real image of the object), the electrons in an SEM do not form a real image.
- SEM scans its electron beam line by line over the sample by using a set of scan coils. This is similar to using a flashlight in a dark room to scan the room from one side to another side.
- Subsequently, the image is built on a monitor (cathode ray tube (CRT)).

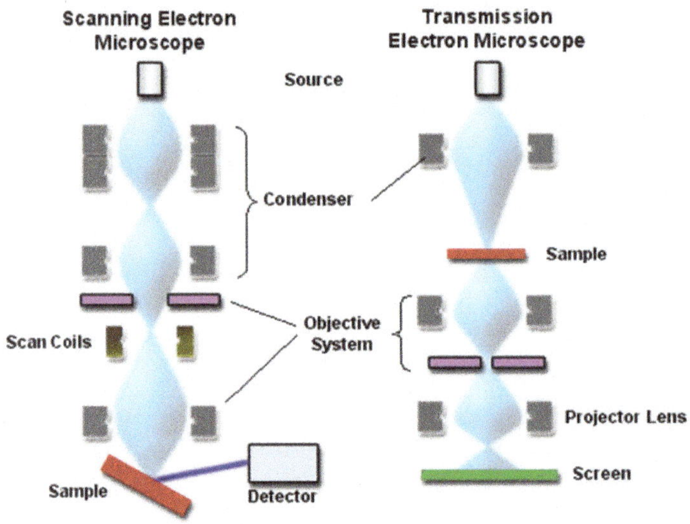

Comparison of ray path diagram between SEM and TEM.

- Figure displays the typical ray path of electrons in SEM and TEM for comparison.

SEM: Interaction with Sample

1. When the electron beam hits a sample surface, it interacts with the atoms in that sample.
2. As a result, some electrons are bounced back out of the sample (backscattered electrons), others knock into atoms and displace electrons that, in turn, come out of the sample (secondary electrons); alternatively X-rays, and light or heat (in the sample) can be the result of these interactions.
3. Generally, the energy is dissipated in the form of heat. We collect the electrons coming out of the material in order to produce the traditional SEM images (called micrographs).
4. The volumes involved in the production of secondary electron, backscattered electron and X-rays, form into a shape (called as Interaction volume) that ranges from a tear-drop to a semi circle within the specimen.
5. The depth and diameter depend on the applied voltage as well as the density of the specimen. The top 15 - 25 nm of the thickness comprises the zone from which secondary electrons can be collected, the top 40% is the region from which backscattered electrons can be collected and X-rays can be collected from the entire region.

Kanaya - Okayama Depth Penetration Formula

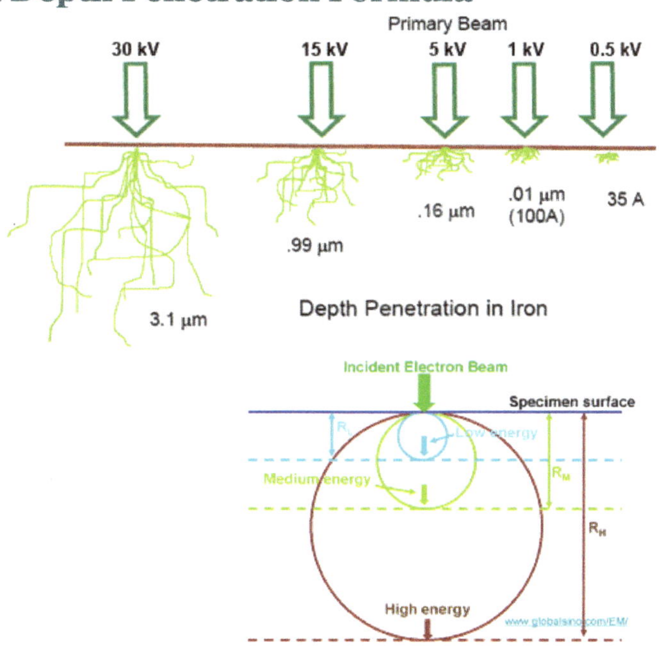

Schematic representation of depth penetration in Iron with beam energies.

The penetration of incident electrons is determined by the electron stopping power of the specimen, which decreases with increasing atomic number (Z). The penetration depth of incident electrons is given by Kanaya -Okayama Formula, as shown in figure.

$$R = \frac{0.0276\ A\ E_0^n}{(Z^{0.89}\ \rho)}\ \mu m$$

R - Depth Penetration

A - Atomic Weight (g/mole)

n - a constant

E_o - Beam Energy (kV)

Z - Atomic number

ρ - Density (g/cm)2

Figure shows the schematic of penetration depth and volume at different beam energies.

- Upon impinging on the specimen, the primary electrons decelerate and in losing energy transfer it inelastically to other atomic electrons and to the lattice. Through continuous random scattering events as shown in figure, the primary beam effectively spreads and fills a teardrop-shaped interaction volume with a multitude of electronic excitations.
- Secondary electrons: Common imaging mode relies on detection of these electrons originating from a subsurface depth of no larger than several angstroms. The signal is captured by a detector consisting of a scintillator -photomultiplier combination, and the output serves to modulate the intensity of a CRT. The image magnification is the ratio of scan lengths on the CRT to that on the specimen.

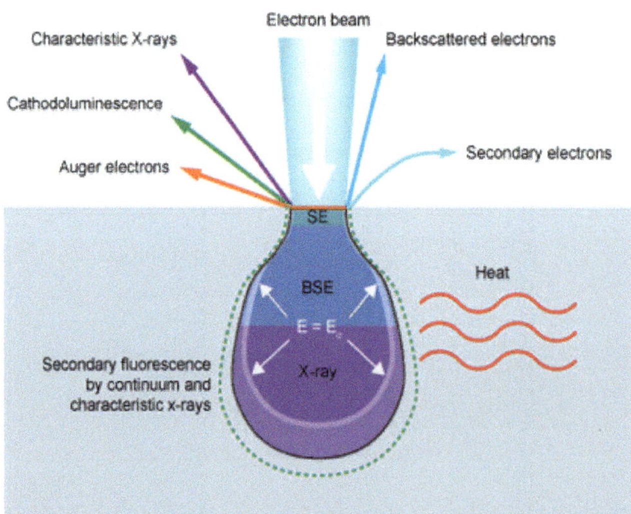

Schematic of electron interacting with specimen surface.

- Backscattered electrons: These are the high energy electrons that are elastically scattered and essentially possess the same energy as the incident electrons. The probability increases with the atomic number Z of the samples. Although the identification of element is not feasible from such information, the contrast can develop between regions of the specimen that differ widely in Z.
- A SEM is like a large X-ray vacuum tube used in conventional X-ray diffraction systems. Electrons emitted from the filament (cathode) are accelerated to high energies where they strike the specimen target (anode). In the process, X-rays characteristics of atoms in the

irradiated area are emitted. By an analysis of their energies, the atoms can be identified and by a count of the numbers of X-rays emitted the concentration of atoms in the specimen can be determined.

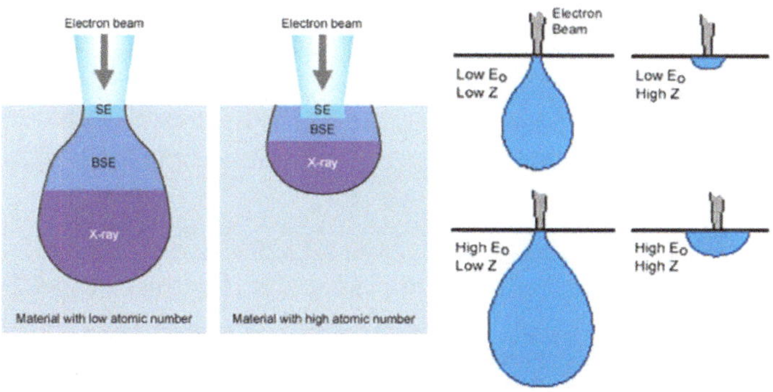

Variation of tear drop shape with accelerating voltage [5,6].

- These figures show the variations in the interaction volume with the atomic number in the materials and energy of the incident beam.
- While the high energy electron beam produces a very nice tear drop shape interaction volume in materials having low atomic number elements, the high atomic number elements produce a very wide interaction volume area which increases with increasing the electron beam energy.

SEM: Images

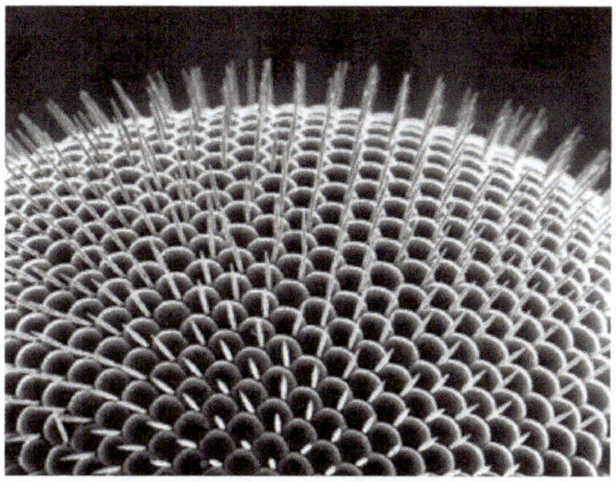

SEM image of an eye on a fruit fly.

Plan-view SEM image of a patterned template for ZnO NR growth.

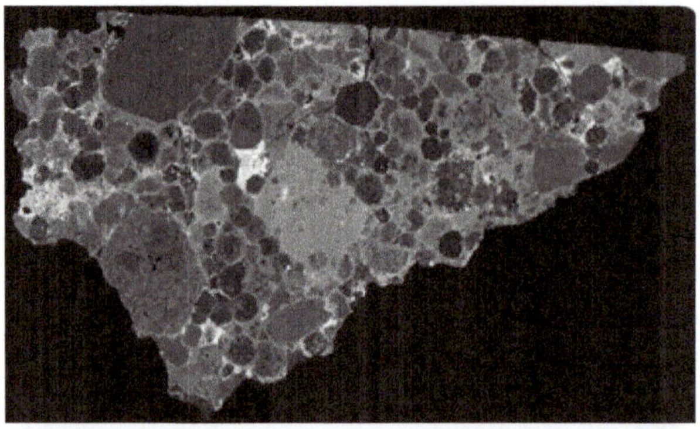

Backscattered image of polished and carbon-coated surfaces of rocks and minerals.

Backscattered image of an aluminum-copper alloy.

SEM: Electron Beam Induced Current

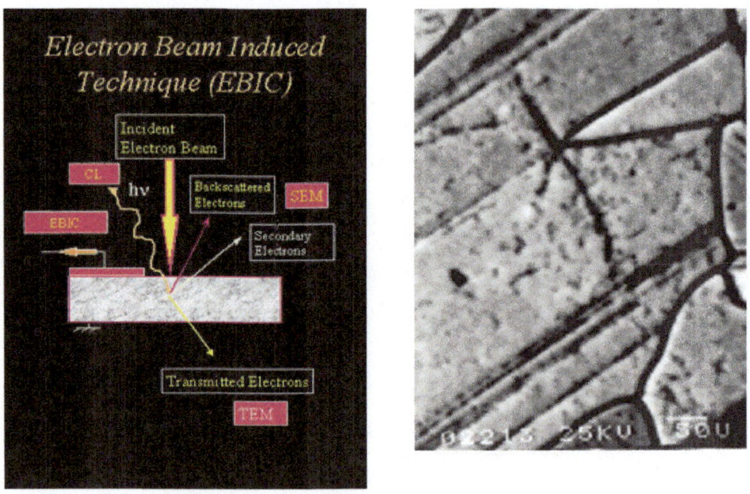

Schematic of EBIC set up and image showing electronically active defects in solar-grade Si.

Electron Beam Induced Current (EBIC) mode is applicable to semiconductor devices. When the primary electrons strike the surface, electron-hole pairs are generated and the resulting current is collected to modulate the intensity of the CRT image.

This technique is useful in spatially locating subsurface defects and failure sites within a junction region.

What are the advantages of FE-SEM over SEM?

SEM	FE-SEM
An electron beam is thermionically emitted from an electron gun fitted with a tungsten filament cathode or LaB_6. Thermionic emitters use electrical current to heat up a filament. When the heat is enough to overcome the work function of the filament, the emission of electrons takes place.	A Field Emission Microscope consists of a metallic sample (tungsten) in the form of a sharp tip with diameter of about 100 nm. The sample is held at a large negative potential (1-10 kV), which gives the electric field near the tip apex to be the order of 10^{10} V/m. This is high enough for field emission of electrons to take place. This is also called as a cold cathode field emitter.
Thermionic sources have the following issues: (1) relatively low brightness, (2) evaporation of cathode material due to heat, and (3) thermal drift during operation.	As a field emission gun emits the electrons from a much smaller area then a thermionic gun, field-emission cathode provides narrower probing beams at low as well as high electron energy. The coherency is much higher and the energy spread is smaller. This further allows the perfectly focused beam.

Rutherford Backscattering Spectrometry

Rutherford backscattering spectrometry (RBS) is an analytical technique used in materials science. Sometimes referred to as high-energy ion scattering (HEIS) spectrometry, RBS is used to determine the structure and composition of materials by measuring the backscattering of a beam of high energy ions (typically protons or alpha particles) impinging on a sample.

The Geiger–Marsden Experiment

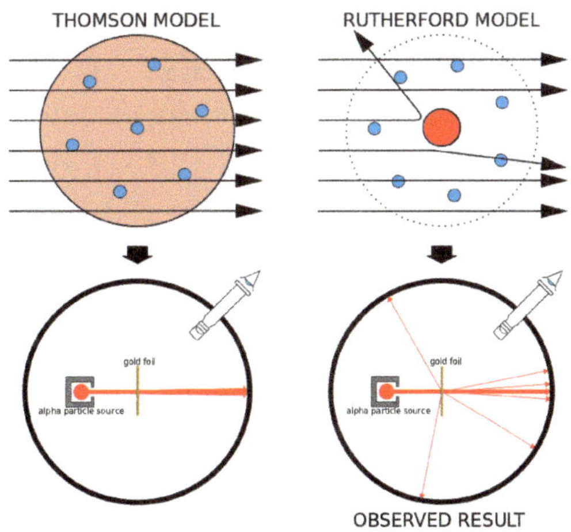

Left: Expected results: alpha particles passing through the plum pudding model of the atom undisturbed.
Right: Observed results: a small portion of the particles were deflected, indicating a small, concentrated positive charge.

Rutherford backscattering spectrometry is named after Lord Ernest Rutherford, a physicist sometimes referred to as the father of nuclear physics. Rutherford supervised a series of experiments carried out by Hans Geiger and Ernest Marsden between 1909 and 1914 studying the scattering of alpha particles through metal foils. While attempting to eliminate "stray particles" they believed to be caused by an imperfection in their alpha source, Rutherford suggested that Marsden attempt to measure backscattering from a gold foil sample. According to the then-dominant plum-pudding model of the atom, in which small negative electrons were spread through a diffuse positive region, backscattering of the high-energy positive alpha particles should have been nonexistent. At most small deflections should occur as the alpha particles passed almost unhindered through the foil. Instead, when Marsden positioned the detector on the same side of the foil as the alpha particle source, he immediately detected a noticeable backscattered signal. According to Rutherford, "It was quite the most incredible event that has ever happened to me in my life. It was almost as incredible as if you fired a 15-inch shell at a piece of tissue paper and it came back and hit you."

Rutherford interpreted the result of the Geiger–Marsden experiment as an indication of a Coulomb collision with a single massive positive particle. This led him to the conclusion that the atom's positive charge could not be diffuse but instead must be concentrated in a single massive core: the atomic nucleus. Calculations indicated that the charge necessary to accomplish this deflection was approximately 100 times the charge of the electron, close to the atomic number of gold. This led to the development of the Rutherford model of the atom in which a positive nucleus made up of *Ne* positive particles, or protons, was surrounded by *N* orbiting electrons of charge -e to balance the nuclear charge. This model was eventually superseded by the Bohr atom, incorporating some early results from quantum mechanics.

If the energy of the incident particle is increased sufficiently, the Coulomb barrier is exceeded and the wavefunctions of the incident and struck particles overlap. This may result in nuclear reactions in certain cases, but frequently the interaction remains elastic, although the scattering cross-sections may fluctuate wildly as a function of energy. This case is known as "Elastic (non-Rutherford) Backscattering Spectrometry" (EBS). There has recently been great progress in determining EBS scattering cross-sections, by solving Schrödinger's equation for each interaction.

Basic Principles

We describe Rutherford backscattering as an elastic, hard-sphere collision between a high kinetic energy particle from the incident beam (the *projectile*) and a stationary particle located in the sample (the *target*). *Elastic* in this context means that no energy is transferred between the incident particle and the stationary particle during the collision, and the state of the stationary particle is not changed. (Except that for a small amount of momentum, which is ignored.)

Note that nuclear interactions are generally not elastic, since a collision may result in a nuclear reaction, with the release of considerable quantities of energy. Nuclear reaction analysis (NRA) is useful for detecting light elements. However, this is not Rutherford scattering.

Considering the kinematics of the collision (that is, the conservation of momentum and kinetic energy), the energy E_1 of the scattered projectile is reduced from the initial energy E_0:

$$E_1 = k \cdot E_0,$$

where k is known as the *kinematical factor*, and

$$k = \left(\frac{m_1 \cos\theta_1 \pm \sqrt{m_2^2 - m_1^2 (\sin\theta_1)^2}}{m_1 + m_2} \right)^2,$$

where particle 1 is the projectile, particle 2 is the target nucleus, and θ_1 is the scattering angle of the projectile in the laboratory frame of reference (that is, relative to the observer). The plus sign is taken when the mass of the projectile is less than that of the target, otherwise the minus sign is taken.

While this equation correctly determines the energy of the scattered projectile for any particular scattering angle (relative to the observer), it does not describe the probability of observing such an event. For that we need the *differential cross-section* of the backscattering event:

$$\frac{d\omega}{d\Omega} = \left(\frac{Z_1 Z_2 e^2}{4E_0}\right)^2 \frac{1}{(\sin\theta/2)^4},$$

where Z_1 and Z_2 are the atomic numbers of the incident and target nuclei. This equation is written in the centre of mass frame of reference and is therefore not a function of the mass of either the projectile or the target nucleus.

Note that the scattering angle in the laboratory frame of reference θ_1 is *not* the same as the scattering angle in the centre of mass frame of reference θ (although for RBS experiments they are usually very similar). However, heavy ion projectiles can easily recoil lighter ions which, if the geometry is right, can be ejected from the target and detected. This is the basis of the Elastic Recoil Detection (ERD, with synonyms ERDA, FRS, HFS) technique. RBS often uses a He beam which readily recoils H, so simultaneous RBS/ERD is frequently done to probe the hydrogen isotope content of samples (although H ERD with a He beam above 1 MeV is not Rutherford. For ERD the scattering angle in the lab frame of reference is quite different from that in the centre of mass frame of reference.

Note also that heavy ions cannot *back*scatter from light ones: it is kinematically prohibited. The kinematical factor must remain real, and this limits the permitted scattering angle in the laboratory frame of reference. In ERD it is often convenient to place the recoil detector at recoil angles large enough to prohibit signal from the scattered beam. The scattered ion intensity is always very large compared to the recoil intensity (the Rutherford scattering cross-section formula goes to infinity as the scattering angle goes to zero), and for ERD the scattered beam usually has to be excluded from the measurement somehow.

The singularity in the Rutherford scattering cross-section formula is unphysical of course. If the scattering cross-section is zero it implies that the projectile never comes close to the target, but in this case it also never penetrates the electron cloud surrounding the nucleus either. The pure Coulomb formula for the scattering cross-section shown above must be corrected for this screening effect, which becomes more important as the energy of the projectile decreases (or, equivalently, its mass increases).

While large-angle scattering only occurs for ions which scatter off target nuclei, inelastic small-angle scattering can also occur off the sample electrons. This results in a gradual decrease in ions which penetrate more deeply into the sample, so that backscattering off interior nuclei occurs with a lower "effective" incident energy. The amount by which the ion energy is lowered after passing through a given distance is referred to as the stopping power of the material and is dependent on the electron distribution. This energy loss varies continuously with respect to distance traversed, so that stopping power is expressed as

$$S(E) = -\frac{dE}{dx}.$$

For high energy ions stopping power is usually proportional to $\frac{Z_2}{E}$; however, precise calculation of stopping power is difficult to carry out with any accuracy.

Stopping power (properly, *stopping force*) has units of energy per unit length. It is generally given in thin film units, that is eV /(atom/cm^2) since it is measured experimentally on thin films whose thickness is always measured absolutely as mass per unit area, avoiding the problem of determining the density of the material which may vary as a function of thickness. Stopping power is now known for all materials at around 2%.

Instrumentation

A single stage 2 MeV linear Van de Graaff particle accelerator, here opened for maintenance

An RBS instrument generally includes three essential components:

- An ion source, usually alpha particles (He^{2+} ions) or, less commonly, protons.
- A linear particle accelerator capable of accelerating incident ions to high energies, usually in the range 1-3 MeV.
- A detector capable of measuring the energies of backscattered ions over some range of angles.

Two common source/acceleration arrangements are used in commercial RBS systems, working in either one or two stages. One-stage systems consist of a He^+ source connected to an acceleration tube with a high positive potential applied to the ion source, and the ground at the end of the acceleration tube. This arrangement is simple and convenient, but it can be difficult to achieve energies of much more than 1 MeV due to the difficulty of applying very high voltages to the system.

Two-stage systems, or "tandem accelerators", start with a source of He^- ions and position the positive terminal at the center of the acceleration tube. A stripper element included in the positive terminal removes electrons from ions which pass through, converting He^- ions to He^{++} ions. The ions thus start out being attracted to the terminal, pass through and become positive, and are repelled until they exit the tube at ground. This arrangement, though more complex, has the advantage of achieving higher accelerations with lower applied voltages: a typical tandem accelerator with an applied voltage of 750 kV can achieve ion energies of over 2 MeV.

Detectors to measure backscattered energy are usually silicon surface barrier detectors, a very thin layer (100 nm) of P-type silicon on an N-type substrate forming a p-n junction. Ions which reach the detector lose some of their energy to inelastic scattering from the electrons, and some of these electrons gain enough energy to overcome the band gap between the semiconductor valence and conduction bands. This means that each ion incident on the detector will produce some number of electron-hole pairs which is dependent on the energy of the ion. These pairs can be detected by applying a voltage across the detector and measuring the current, providing an effective measurement of the ion energy. The relationship between ion energy and the number of electron-hole pairs produced will be dependent on the detector materials, the type of ion and the efficiency of the current measurement; energy resolution is dependent on thermal fluctuations. It should also be noted that after one ion is incident on the detector, there will be some *dead time* before the electron-hole pairs recombine in which a second incident ion cannot be distinguished from the first.

Angular dependence of detection can be achieved by using a movable detector, or more practically by separating the surface barrier detector into many independent cells which can be measured independently, covering some range of angles around direct (180 degrees) back-scattering. Angular dependence of the incident beam is controlled by using a tiltable sample stage.

Composition and Depth Measurement

The energy loss of a backscattered ion is dependent on two processes: the energy lost in scattering events with sample nuclei, and the energy lost to small-angle scattering from the sample electrons. The first process is dependent on the scattering cross-section of the nucleus and thus on its mass and atomic number. For a given measurement angle, nuclei of two different elements will therefore scatter incident ions to different degrees and with different energies, producing separate peaks on an N(E) plot of measurement count versus energy. These peaks are characteristic of the elements contained in the material, providing a means of analyzing the composition of a sample by matching scattered energies to known scattering cross-sections. Relative concentrations can be determined by measuring the heights of the peaks.

The second energy loss process, the stopping power of the sample electrons, does not result in large discrete losses such as those produced by nuclear collisions. Instead it creates a gradual energy loss dependent on the electron density and the distance traversed in the sample. This energy loss will lower the measured energy of ions which backscatter from nuclei inside the sample in a continuous manner dependent on the depth of the nuclei. The result is that instead of the sharp backscattered peaks one would expect on an N(E) plot, with the width determined by energy and angular resolution, the peaks observed trail off gradually towards lower energy as the ions pass through the depth occupied by that element. Elements which only appear at some depth inside the sample will also have their peak positions shifted by some amount which represents the distance an ion had to traverse to reach those nuclei.

In practice, then, a compositional depth profile can be determined from an RBS N(E) measurement. The elements contained by a sample can be determined from the positions of peaks in the energy spectrum. Depth can be determined from the width and shifted position of these peaks, and relative concentration from the peak heights. This is especially useful for the analysis of a multilayer sample, for example, or for a sample with a composition which varies more continuously with depth.

This kind of measurement can only be used to determine elemental composition; the chemical structure of the sample cannot be determined from the N(E) profile. However, it is possible to learn something about this through RBS by examining the crystal structure. This kind of spatial information can be investigated by taking advantage of blocking and channeling.

Structural Measurements: Blocking and Channeling

To fully understand the interaction of an incident beam of nuclei with a crystalline structure, it is necessary to comprehend two more key concepts: *blocking* and *channeling*.

When a beam of ions with parallel trajectories is incident on a target atom, scattering off that atom will prevent collisions in a cone-shaped region "behind" the target relative to the beam. This occurs because the repulsive potential of the target atom bends close ion trajectories away from their original path, and is referred to as *blocking*. The radius of this blocked region, at a distance L from the original atom, is given by

$$R = 2\sqrt{\frac{Z_1 Z_2 e^2 L}{E_0}}$$

When an ion is scattered from deep inside a sample, it can then re-scatter off a second atom, creating a second blocked cone in the direction of the scattered trajectory. This can be detected by carefully varying the detection angle relative to the incident angle.

Channeling is observed when the incident beam is aligned with a major symmetry axis of the crystal. Incident nuclei which avoid collisions with surface atoms are excluded from collisions with all atoms deeper in the sample, due to blocking by the first layer of atoms. When the interatomic distance is large compared to the radius of the blocked cone, the incident ions can penetrate many times the interatomic distance without being backscattered. This can result in a drastic reduction of the observed backscattered signal when the incident beam is oriented along one of the symmetry directions, allowing determination of a sample's regular crystal structure. Channeling works best for very small blocking radii, i.e. for high-energy, low-atomic-number incident ions such as He^+.

The tolerance for the deviation of the ion beam angle of incidence relative to the symmetry direction depends on the blocking radius, making the allowable deviation angle proportional to

$$\sqrt{\frac{Z_1 Z_2}{E_0 d}}$$

While the intensity of an RBS peak is observed to decrease across most of its width when the beam is channeled, a narrow peak at the high-energy end of larger peak will often be observed, representing surface scattering from the first layer of atoms. The presence of this peak opens the possibility of surface sensitivity for RBS measurements.

Profiling of Displaced Atoms

In addition, channeling of ions can also be used to analyze a crystalline sample for lattice damage. If atoms within the target are displaced from their crystalline lattice site, this will result in a higher backscattering yield in relation to a perfect crystal. By comparing the spectrum from a sample being analyzed to that from a perfect crystal, and that obtained at a random (non-channeling) orientation (representative of a spectrum from an amorphous sample), it is possible to determine the extent of crystalline damage in terms of a fraction of displaced atoms. Multiplying this fraction by the density of the material when amorphous then also gives an estimate for the concentration of displaced atoms. The energy at which the increased backscattering occurs can also be used to determine the depth at which the displaced atoms are and a defect depth profile can be built up as a result.

Surface Sensitivity

While RBS is generally used to measure the bulk composition and structure of a sample, it is possible to obtain some information about the structure and composition of the sample surface. When the signal is channeled to remove the bulk signal, careful manipulation of the incident and detection angles can be used to determine the relative positions of the first few layers of atoms, taking advantage of blocking effects.

The surface structure of a sample can be changed from the ideal in a number of ways. The first layer of atoms can change its distance from subsequent layers (relaxation); it can assume a different two-dimensional structure than the bulk (reconstruction); or another material can be adsorbed onto the surface. Each of these cases can be detected by RBS. For example, surface reconstruction can be detected by aligning the beam in such a way that channeling should occur, so that only a surface peak of known intensity should be detected. A higher-than-usual intensity or a wider peak will indicate that the first layers of atoms are failing to block the layers beneath, i.e. that the surface has been reconstructed. Relaxations can be detected by a similar procedure with the sample tilted so the ion beam is incident at an angle selected so that first-layer atoms should block backscattering at a diagonal; that is, from atoms which are below and displaced from the blocking atom. A higher-than-expected backscattered yield will indicate that the first layer has been displaced relative to the second layer, or relaxed. Adsorbate materials will be detected by their different composition, changing the position of the surface peak relative to the expected position.

RBS has also been used to measure processes which affect the surface differently from the bulk by analyzing changes in the channeled surface peak. A well-known example of this is the RBS analysis of the premelting of lead surfaces by Frenken, Maree and van der Veen. In an RBS measurement of the Pb(110) surface, a well-defined surface peak which is stable at low temperatures was found to become wider and more intense as temperature increase past two-thirds of the bulk melting temperature. The peak reached the bulk height and width as temperature reached the melting temperature. This increase in the disorder of the surface, making deeper atoms visible to the incident beam, was interpreted as pre-melting of the surface, and computer simulations of the RBS process produced similar results when compared with theoretical pre-melting predictions.

RBS has also been combined with nuclear microscopy, in which a focused ion beam is scanned across a surface in a manner similar to a scanning electron microscope. The energetic analysis of backscattered signals in this kind of application provides compositional information about the surface, while the microprobe itself can be used to examine features such as periodic surface structures.

Rutherford Back Scattering

RBS is based on collisions between atomic nuclei and involves measuring the number and energy of ions in a beam which backscatter after colliding with atoms in the near-surface region of a sample at which the beam has been targeted. It derives its name from Lord Ernest Rutherford, who in 1911 was the first to present the concept of atoms having nuclei.

Photoview of Lord Ernest Rutherford.

- RBS or high-energy ion scattering (HEIS) spectrometry is an analytical technique used in materials science. This technique is used to determine the structure and composition of materials by measuring the backscattering of a beam of high energy ions impinging on a sample.
- Elemental analysis technique where alpha particles are scattered off by the targets and is used to determine elemental composition and thin-film thickness.

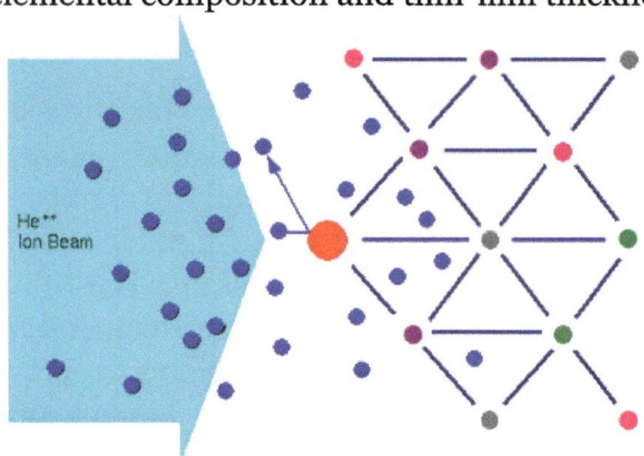

Schematic of interaction of He ion beam with materials.

- RBS is an ion scattering technique that is used for the surface layer analysis of solids.
- A target is bombarded by the ions with energies in the range of 0.5 - 4 MeV.

- The energy of the backscattered projectiles is recorded with an energy sensitive detector, typically a solid state detector.
- RBS allows the quantitative determination of the composition of a material and depth profiling of individual elements.
- It does not need any reference samples and a nondestructive technique.
- It has a good depth resolution of the order of several nm.
- RBS is very sensitive for heavy elements of the order of parts-per-million (ppm).

The major points to learn are:

1. Scattering is proportional to the squares of both the charge of the incident particle and the target nucleus.
2. The number of scattered particles varies as the inverse square of the kinetic energy of the incident particle.
3. Scattering is inversely proportional to the fourth power of sin(θ/2), where θ is the scattering angle.
4. For thin targets, the scattering is proportional to the target thickness, t.

- Consider an ion of charge z, mass m and initial kinetic energy Eo being scattered elastically from a stationary nucleus of charge Z and mass M purely by the Coulomb force. The final energy E of the scattered ion is a function of the angle of scatter θ from the initial direction and the ratio M/m.

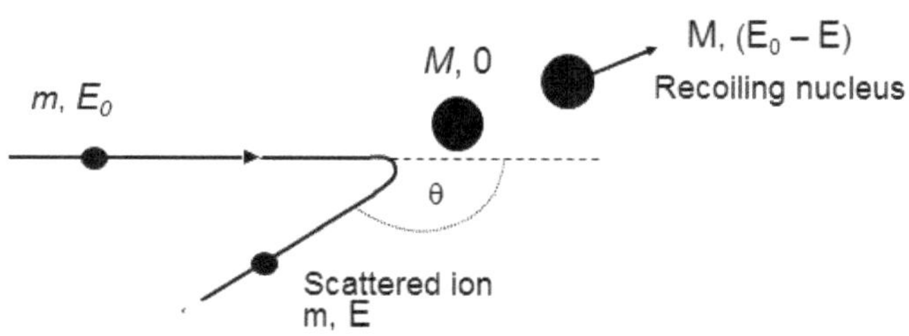

Schematic of interaction between ion and nuclei.

- The scattered E can be derived from the principles of conservation of energy and momentum, and is given by (in laboratory frame of reference):

$$E = \left[\left(\frac{1}{1+\frac{M}{m}}\right) \times \left(\cos\theta + \sqrt{\left(\frac{M}{m}\right)^2 - \sin^2\theta}\right)\right]^2 E_0 = kE_0$$

- The multiplication factor k on the right hand side of the equation is often referred to as the kinematic factor, which is the ratio of the scattered alpha particle's energy to its incident energy gives a convenient method for determining the elemental make up of the target atoms in the sample. For M/m > 1, k is a slow-varying function of θ, having the maximum value of 1 at θ = 0 and the minimum value at θ = 180°. For M/m = 1, the value of k is zero beyond 90°. The plots of k versus θ and k versus Eo are displayed below.

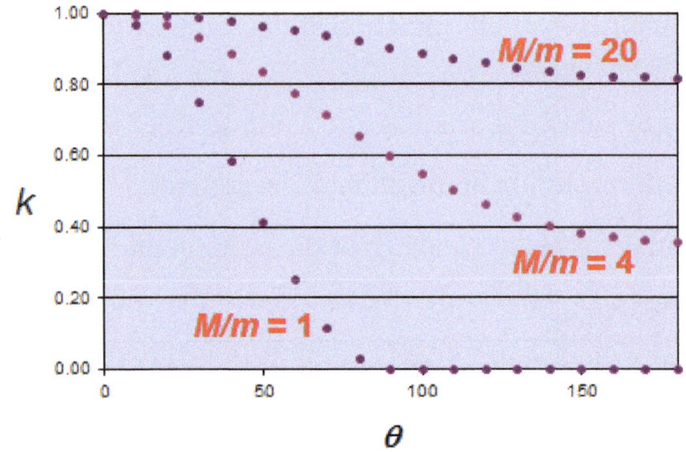

The plot of *k* versus θ for *M/m* = 0, 4, 20. *k* values depend strongly on the ratio between *M/m*.

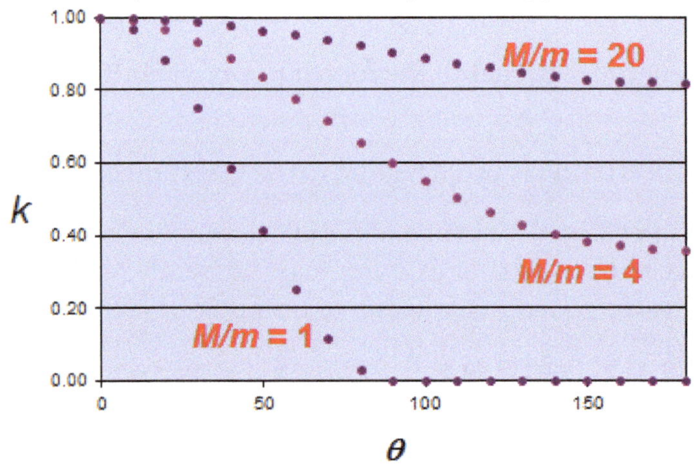

The plot of *k* versus θ = 180°.

RBS: Differential Scattering Cross Section

The theoretical differential cross section for a given scattering angle is given by:

$$\frac{d\sigma}{d\Omega}=\left(\frac{zZe^2}{4E_0}\right)^2\times\left[\sin^{-4}\left(\frac{\theta}{2}\right)-2\left(\frac{m}{M}\right)^2+\ldots\ldots\ldots\right]$$

The higher order terms are usually negligible for M>m and the differential cross section can be expressed in the following form, which has the unit of mb / sr (millibarns per steradian) when MeV is used as the unit for E_0:

$$\frac{d\sigma}{d\Omega}=1.296\times\left(\frac{zZ}{E_0}\right)^2\times\left[\sin^{-4}\left(\frac{\theta}{2}\right)-2\left(\frac{m}{M}\right)^2\right]$$

The variation of dσ/dΩ as a function of scattering angle and incident ion energy are described.

dσ/dΩ versus *θ* :

The dependence of dσ/dΩ on sin-4(θ /2) means that dσ/dΩ has very large values at small forward scattered angles and approaches infinity at zero degree. However, for θ = 90° - 180°, it decreases slowly because sin-4 (θ /2) drops gradually from a value of 4 at 90 o to 1 at 180°. Figure depicts the plots of dσ/dΩ vs θ for the Coulomb scattering of 2 MeV 4He+ ions for oxygen and calcium.

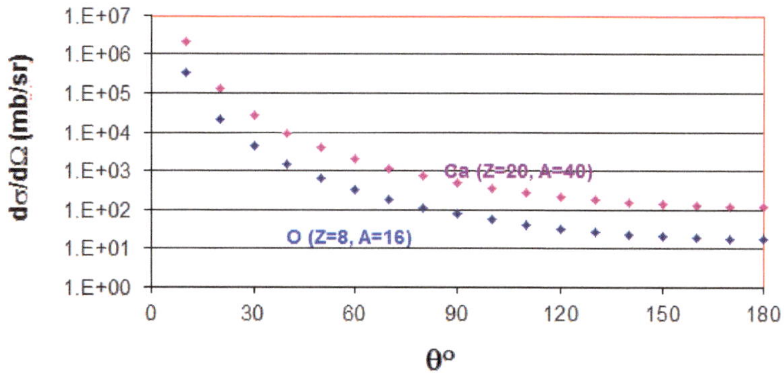

The variation of dσ/dΩ versus θ.

dσ/dΩ versus E_o :

dσ/dΩ is inversely proportional to the square of the incident ion energy E_0, implying that it increases with decreasing E_0. This also means that for a thick target, more ions are scattered from a greater depth. Figure shows a plot of dσ/dΩ vs E_0 for the scattering of 4He+ from an aluminum target at θ = 180°.

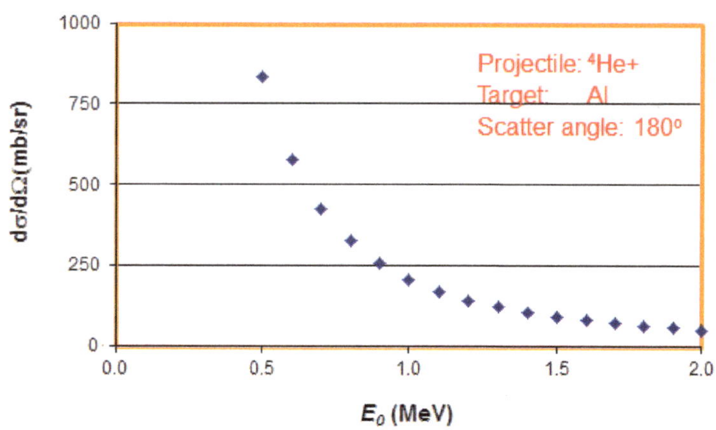

The variation of dσ/dΩ versus E_0.

RBS: Yield for Thin Target Materials

The number of backscattered ions detected by a particle detector at an angle of θ and subtending a solid angle ΔΩ at the target is called the yield. Its relation with the differential cross section is as follows:

$$Y(\theta) = n_0 n_z \frac{d\sigma}{d\Omega} \Delta\Omega$$

where no is the number of incident ions and nz is the areal concentration of the target atom. This is only valid for thin targets in which the loss of energy by an incident proton in them is negligibly small.

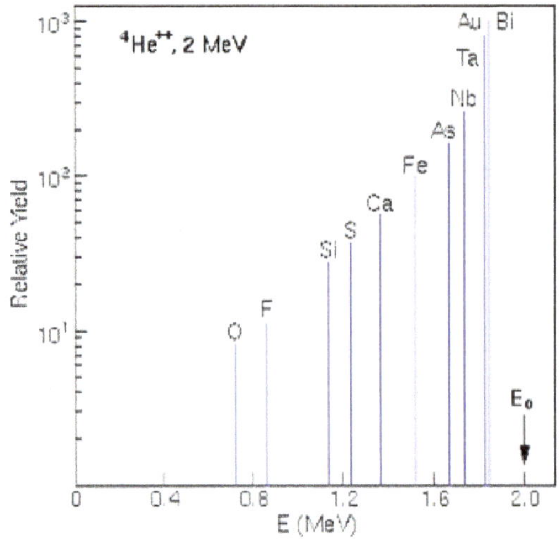

The variation of relative field versus E.

Figure displays relative yields for He backscattering from selected elements at an incident He energy of 2MeV. The energies for He backscattering from these elements when present at the surface of a sample are also displayed. This reveals that RBS is more sensitive for heavy elements than for light elements, due to the larger scattering cross sections of the heavier elements.

RBS: Experiment

- One of the well-suited samples for RBS analysis is a metal silicide film. The conductivity depends on the ratio of the silicon to the metal and on the thickness of the film. Both are easily determined by RBS.

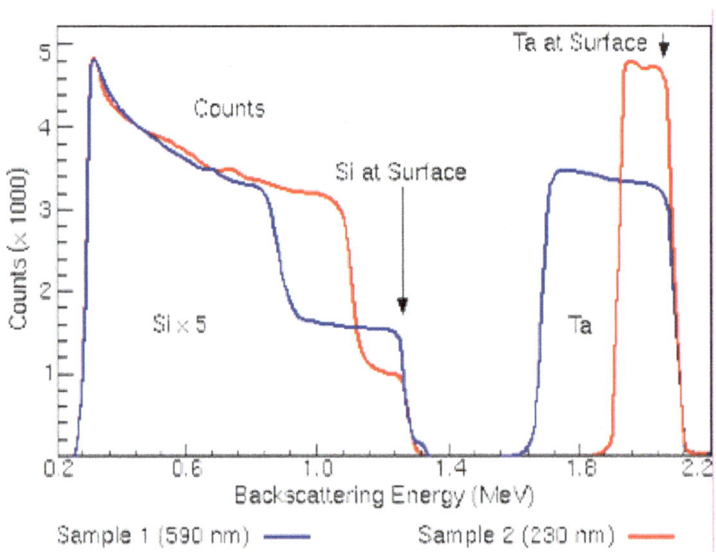

The plot of counts versus backscattering energy.

- Figure illustrates RBS spectra from two samples: one is 230 nm thick and the other is 590 nm thick. The experiment uses an ion beam of He++ at 2.2 MeV. In both spectra, the high energy peak arises by scattering from tantalum in the TaSi film layer. The peak at lower energy is from silicon, which appears in both the TaSi film on the surface and in the Si substrate.

- Particles scattered from tantalum at the TaSi /Si interface of the 230 nm film have a final energy of about 1.9 MeV, while particles scattered from the same interface of the 590 nm film have less final energy (about 1.7 MeV) because they have passed through more TaSi. The entire Ta peak spans a greater energy range, because of the increased thickness of the layer it represents.
- The typical thickness calculation using RBS depth profile is shown in figure.

RBS Experiment: Thickness Determination

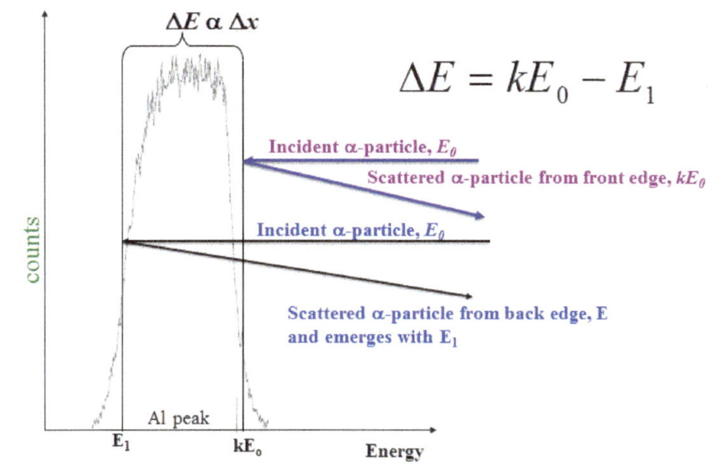

Schematic of calculating thickness from RBS data.

RBS Experiment: Thickness Determination

$$\text{Inward}: \frac{\Delta x}{\cos\theta_1} = \Delta x = -\int_{E_o}^{E} dE \Big/ \left(\frac{\partial E}{\partial x}\right)$$

θ_1 - angle between incident beam and horizontal.

$$\text{Outward}: \frac{\Delta x}{\cos\theta_2} = \Delta x = -\int_{kE}^{E_1} dE \Big/ \left(\frac{\partial E}{\partial x}\right)$$

θ_2 - angle between exiting beam and horizontal.

If $\partial E / \partial x$ is assumed constant along the inward and outward paths, then

$$\text{Inward}: \Delta x = -\int_{E_o}^{E} dE \Big/ \left(\frac{\partial E}{\partial x}\right) = -\left.\frac{\partial x}{\partial E}\right|_{in} (E - E_0)$$

$$\text{Outward}: \frac{\Delta x}{\cos\theta_2} = -\int_{kE}^{E_1} dE \Big/ \left(\frac{\partial E}{\partial x}\right) = -\left.\frac{\partial x}{\partial E}\right|_{out} (E_1 - kE)$$

Here, we don't know the energy at the back edge. So we can eliminate this energy E.

RBS Experiment: Effect of Interlayer

RBS spectra of the Fe/SmCo bilayers deposited on the Si (100) substrate are shown in figure. The solid curves represent the simulated spectra, while the open circles show the experimental data. The simulation was performed using the SIMNRA software, which allows the determination of thickness and composition of each individual layer.

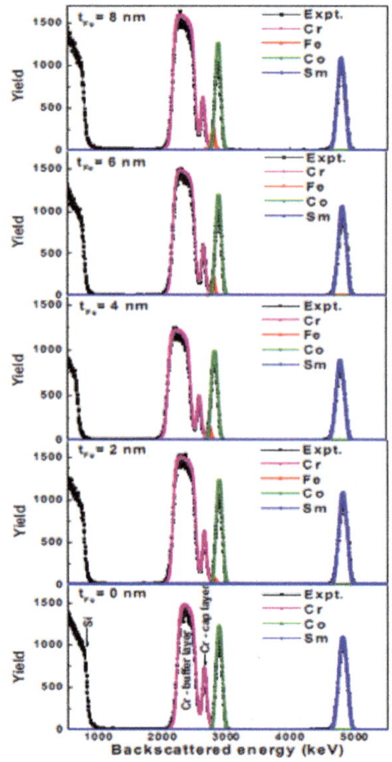

RBS spectra of the Cr(10nm)/ Fe(t nm)/ Sm-Co (20 nm)/ Cr (90nm) bilayers with different Fe thicknesses.

A monotonic increase in the intensity of Fe component can be observed with increasing nominal Fe layer thickness. More accurate evaluation of the composition and thickness of the diffused Fe(-Co) layer by RBS is somewhat limited due to the overlapping of signals from elements such as Cr, Fe and Co because of their very similar masses. It should be noted that RBS results show that diffusion of Fe into the Sm-Co layer plays a minor role since related models did not result in better fits.

The calculated values of Fe layer thicknesses and diffused Co content in the Fe layer are presented in table.

Fe nominal thickness (nm)	Cr (buffer layer) t_{Cr} (nm)	Sm–Co (hard layer)		Fe (soft layer)		Cr (cap layer) t_{Cr} (nm)
		C (at%)	t_{SmCo} (nm)	C (at%)	t_{Fe} (nm)	
0	88.9	Sm: 17 Co: 83	21.6	—	—	10.6
2	87.4	Sm: 18 Co: 82	20.3	Fe: 53 Co: 47	1.6	9.5
4	91.6	Sm: 17 Co: 83	20.7	Fe: 68 Co: 32	3.7	9.3
6	87.3	Sm: 17 Co: 83	19.3	Fe: 74 Co: 26	5.6	8.9
8	86.7	Sm: 18 Co: 82	20.1	Fe: 81 Co: 19	7.2	9.7

Note: C is the atomic composition.

Atomic-force Microscopy

Atomic-force microscopy (AFM) or scanning-force Microscopy (SFM) is a very-high-resolution type of scanning probe microscopy (SPM), with demonstrated resolution on the order of fractions of a nanometer, more than 1000 times better than the optical diffraction limit.

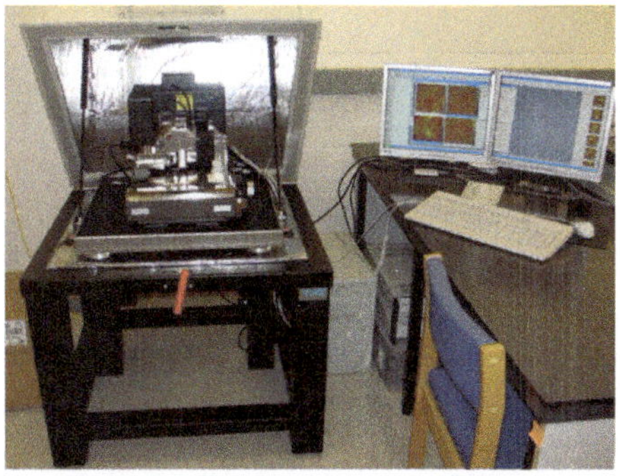

An atomic-force microscope on the left with controlling computer on the right.

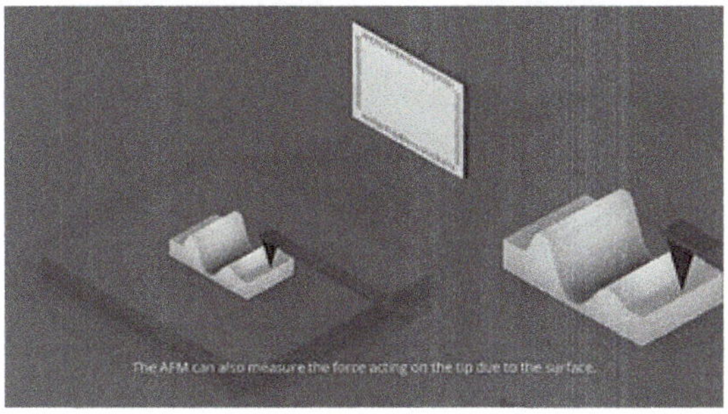

Atomic Force Microscope

Overview

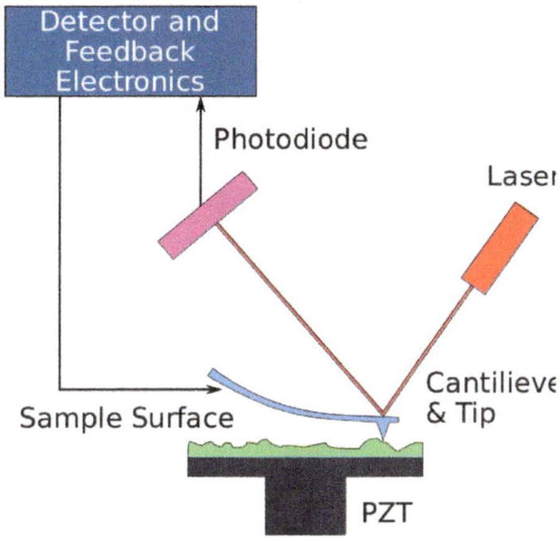

Block diagram of atomic-force microscope using beam deflection detection. As the cantilever is displaced via its interaction with the surface, so too will the reflection of the laser beam be displaced on the surface of the photodiode.

Atomic-force microscopy (AFM) or scanning-force microscopy (SFM) is a type of scanning probe microscopy (SPM), with demonstrated resolution on the order of fractions of a nanometer, more than 1000 times better than the optical diffraction limit. The information is gathered by "feeling" or "touching" the surface with a mechanical probe. Piezoelectric elements that facilitate tiny but accurate and precise movements on (electronic) command enable very precise scanning.

Abilities

The AFM has three major abilities: force measurement, imaging, and manipulation.

In force measurement, AFMs can be used to measure the forces between the probe and the sample as a function of their mutual separation. This can be applied to perform force spectroscopy.

For imaging, the reaction of the probe to the forces that the sample imposes on it can be used to form an image of the three-dimensional shape (topography) of a sample surface at a high resolution. This is achieved by raster scanning the position of the sample with respect to the tip and recording the height of the probe that corresponds to a constant probe-sample interaction. The surface topography is commonly displayed as a pseudocolor plot.

In manipulation, the forces between tip and sample can also be used to change the properties of the sample in a controlled way. Examples of this include atomic manipulation, scanning probe lithography and local stimulation of cells.

Simultaneous with the acquisition of topographical images, other properties of the sample can be measured locally and displayed as an image, often with similarly high resolution. Examples of such properties are mechanical properties like stiffness or adhesion strength and electrical properties such as conductivity or surface potential. In fact, the majority of SPM techniques are extensions of AFM that use this modality.

Other Microscopy Technologies

Compared to competitive technologies such as optical microscopy and electron microscopy, the major difference between these and the atomic-force microscope is that the latter does not use lenses or beam irradiation. Therefore, it does not suffer from a limitation of space resolution due to diffraction limit and aberration, and it is not necessary to prepare a space for guiding the beam (by creating a vacuum) or to stain the sample.

There are several types of scanning microscopy including scanning probe microscopy (which includes AFM, scanning tunneling microscopy (STM) and near-field scanning optical microscope (SNOM/NSOM), STED microscopy (STED), and scanning electron microscopy). Although SNOM and STED use visible light to illuminate the sample, their resolution is not constrained by the diffraction limit.

Configuration

Figure shows an AFM typically consisting of the following features:

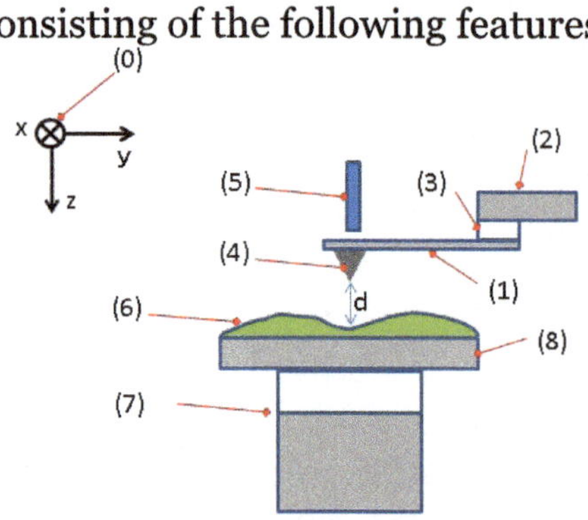

Typical configuration of an AFM. (1):: Cantilever, (2): Support for cantilever, (3): Piezoelectric element(to oscillate cantilever at its eigen frequency.), (4): Tip (Fixed to open end of a cantilever, acts as the probe), (5): Detector of deflection and motion of the cantilever, (6): Sample to be measured by AFM, (7): xyz drive, (moves sample (6) and stage (8) in x, y, and z directions with respect to a tip apex (4)), and (8): Stage.

The small spring-like cantilever (1) is carried by the support (2). Optionally, a piezoelectric element (3) oscillates the cantilever (1). The sharp tip (4) is fixed to the free end of the cantilever (1). The detector (5) records the deflection and motion of the cantilever (1). The sample (6) is mounted on the sample stage (8). An xyz drive (7) permits to displace the sample (6) and the sample stage (8) in x, y, and z directions with respect to the tip apex (4). Although figure shows the drive attached to the sample, the drive can also be attached to the tip, or independent drives can be attached to both, since it is the relative displacement of the sample and tip that needs to be controlled. Controllers and plotter are not shown in figure. Numbers in parentheses correspond to numbered features in figure. Coordinate directions are defined by the coordinate system (0).

According to the configuration described above, the interaction between tip and sample, which can be an atomic scale phenomenon, is transduced into changes of the motion of cantilever which is a macro scale phenomenon. Several different aspects of the cantilever motion can be used to quantify the interaction between the tip and sample, most commonly the value of the deflection, the amplitude of an imposed oscillation of the cantilever, or the shift in resonance frequency of the cantilever.

Detector

The detector (5) of AFM measures the deflection (displacement with respect to the equilibrium position) of the cantilever and converts it into an electrical signal. The intensity of this signal will be proportional to the displacement of the cantilever.

Various methods of detection can be used, e.g. interferometry, optical levers, the piezoresistive method, the piezoelectric method, and STM-based detectors.

Image Formation

When using the AFM to image a sample, the tip is brought into contact with the sample, and the sample is raster scanned along an x-y grid figure. Most commonly, an electronic feedback loop is employed to keep the probe-sample force constant during scanning. This feedback loop has the cantilever deflection as input, and its output controls the distance along the z axis between the probe support figure and the sample support figure. As long as the tip remains in contact with the sample, and the sample is scanned in the x-y plane, height variations in the sample will change the deflection of the cantilever. The feedback then adjusts the height of the probe support so that the deflection is restored to a user-defined value (the setpoint). A properly adjusted feedback loop adjusts the support-sample separation continuously during the scanning motion, such that the deflection remains approximately constant. In this situation, the feedback output equals the sample surface topography to within a small error.

Historically, a different operation method has been used, in which the sample-probe support distance is kept constant and not controlled by a feedback (servo mechanism). In this mode, usually referred to as 'constant height mode', the deflection of the cantilever is recorded as a function of the sample x-y position. As long as the tip is in contact with the sample, the deflection then corresponds to surface topography. The main reason this method is not very popular anymore, is that the forces between tip and sample are not controlled, which can lead to forces high enough to damage the tip or the sample. It is however common practice to record the deflection even when scanning in 'constant force mode', with feedback. This reveals the small tracking error of the feedback, and can sometimes reveal features that the feedback was not able to adjust for.

The AFM signals, such as sample height or cantilever deflection, are recorded on a computer during the x-y scan. They are plotted in a pseudocolor image, in which each pixel represents an x-y position on the sample, and the color represents the recorded signal.

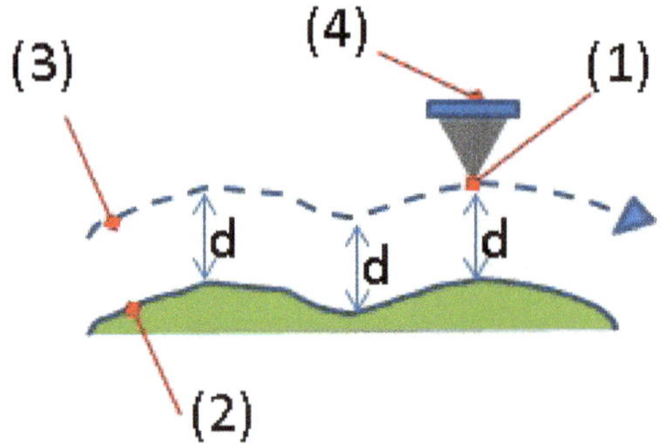

Topographic image forming by AFM.
(1): Tip apex, (2): Sample surface, (3): Z-orbit of Tip apex, (4): Cantilever.

History

AFM was invented by IBM Scientists in 1982.The precursor to the AFM, the scanning tunneling microscope (STM), was developed by Gerd Binnig and Heinrich Rohrer in the early 1980s at IBM Research - Zurich, a development that earned them the Nobel Prize for Physics in 1986. Binnig invented the atomic-force microscope and the first experimental implementation was made by Binnig, Quate and Gerber in 1986.

The first commercially available atomic-force microscope was introduced in 1989. The AFM is one of the foremost tools for imaging, measuring, and manipulating matter at the nanoscale.

Applications

The AFM has been applied to problems in a wide range of disciplines of the natural sciences, including solid-state physics, semiconductor science and technology, molecular engineering, polymer chemistry and physics, surface chemistry, molecular biology, cell biology and medicine.

Applications in the field of solid state physics include (a) the identification of atoms at a surface, (b) the evaluation of interactions between a specific atom and its neighboring atoms, and (c) the study of changes in physical properties arising from changes in an atomic arrangement through atomic manipulation.

In cellular biology, AFM can be used to (a) attempt to distinguish cancer cells and normal cells based on a hardness of cells, and (b) to evaluate interactions between a specific cell and its neighboring cells in a competitive culture system.

In some variations, electric potentials can also be scanned using conducting cantilevers. In more advanced versions, currents can be passed through the tip to probe the electrical conductivity or transport of the underlying surface, but this is a challenging task with few research groups reporting consistent data (as of 2004).

Principles

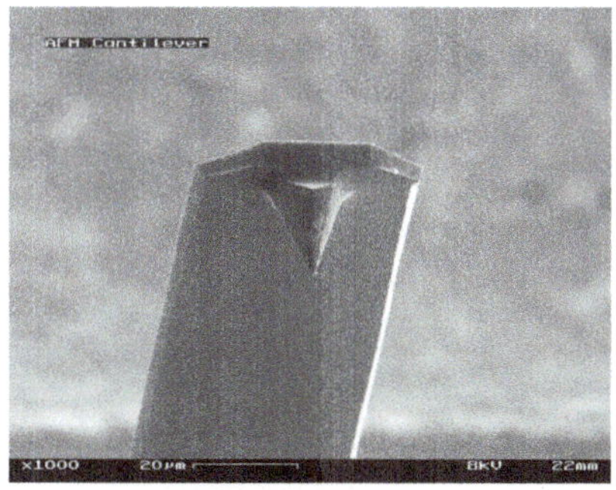

Electron micrograph of a used AFM cantilever. Image width ~100 micrometers

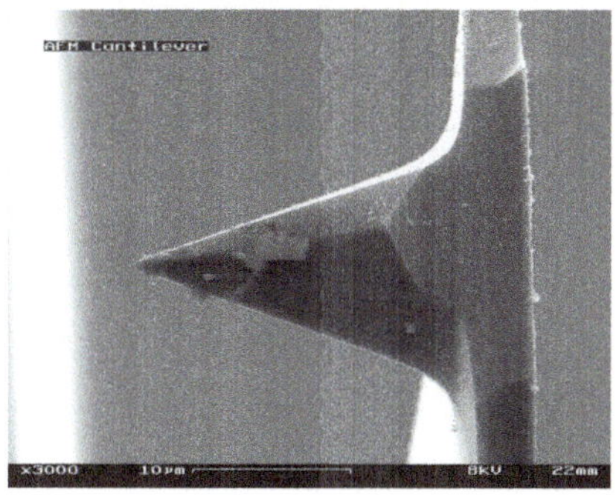

Electron micrograph of a used AFM cantilever. Image width ~30 micrometers

The AFM consists of a cantilever with a sharp tip (probe) at its end that is used to scan the specimen surface. The cantilever is typically silicon or silicon nitride with a tip radius of curvature on the order of nanometers. When the tip is brought into proximity of a sample surface, forces between the tip and the sample lead to a deflection of the cantilever according to Hooke's law. Depending on the situation, forces that are measured in AFM include mechanical contact force, van der Waals forces, capillary forces, chemical bonding, electrostatic forces, magnetic forces, Casimir forces, solvation forces, etc. Along with force, additional quantities may simultaneously be measured through the use of specialized types of probes.

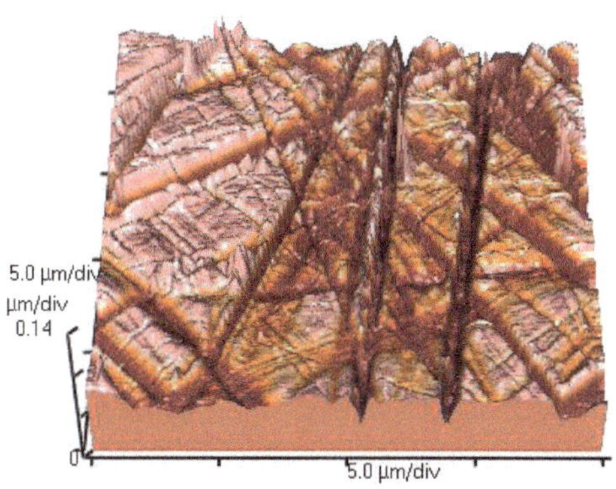

Atomic-force microscope topographical scan of a glass surface. The micro and nano-scale features of the glass can be observed, portraying the roughness of the material. The image space is (x,y,z) = (20 μm × 20 μm × 420 nm).

The AFM can be operated in a number of modes, depending on the application. In general, possible imaging modes are divided into static (also called *contact*) modes and a variety of dynamic (non-contact or "tapping") modes where the cantilever is vibrated or oscillated at a given frequency.

Imaging Modes

AFM operation is usually described as one of three modes, according to the nature of the tip motion: contact mode, also called static mode (as opposed to the other two modes, which are called dynamic modes); tapping mode, also called intermittent contact, AC mode, or vibrating mode, or, after the detection mechanism, amplitude modulation AFM; non-contact mode, or, again after the detection mechanism, frequency modulation AFM.

It should be noted that despite the nomenclature, repulsive contact can occur or be avoided both in amplitude modulation AFM and frequency modulation AFM, depending on the settings.

Contact Mode

In contact mode, the tip is "dragged" across the surface of the sample and the contours of the surface are measured either using the deflection of the cantilever directly or, more commonly, using the feedback signal required to keep the cantilever at a constant position. Because the measurement of a static signal is prone to noise and drift, low stiffness cantilevers (i.e. cantilevers with a low spring constant, k) are used to achieve a large enough deflection signal while keeping the interaction force low. Close to the surface of the sample, attractive forces can be quite strong, causing the tip to "snap-in" to the surface. Thus, contact mode AFM is almost always done at a depth where the overall force is repulsive, that is, in firm "contact" with the solid surface.

Tapping Mode

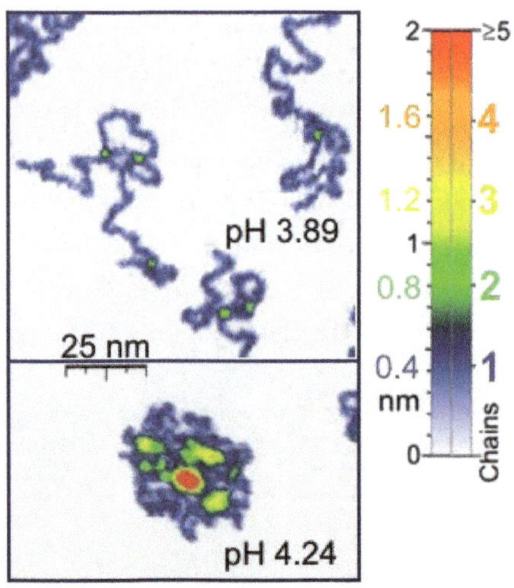

Single polymer chains (0.4 nm thick) recorded in a tapping mode under aqueous media with different pH.

In ambient conditions, most samples develop a liquid meniscus layer. Because of this, keeping the probe tip close enough to the sample for short-range forces to become detectable while preventing the tip from sticking to the surface presents a major problem for contact mode in ambient conditions. Dynamic contact mode (also called intermittent contact, AC mode or tapping mode) was developed to bypass this problem. Nowadays, tapping mode is the most frequently used AFM mode when operating in ambient conditions or in liquids.

In *tapping mode*, the cantilever is driven to oscillate up and down at or near its resonance frequency. This oscillation is commonly achieved with a small piezo element in the cantilever holder, but other possibilities include an AC magnetic field (with magnetic cantilevers), piezoelectric cantilevers, or periodic heating with a modulated laser beam. The amplitude of this oscillation usually varies from several nm to 200 nm. In tapping mode, the frequency and amplitude of the driving signal are kept constant, leading to a constant amplitude of the cantilever oscillation as long as there is no drift or interaction with the surface. The interaction of forces acting on the cantilever when the tip comes close to the surface, Van der Waals forces, dipole-dipole interactions, electrostatic forces, etc. cause the amplitude of the cantilever's oscillation to change (usually decrease) as the tip gets closer to the sample. This amplitude is used as the parameter that goes into the electronic servo that controls the height of the cantilever above the sample. The servo adjusts the height to maintain a set cantilever oscillation amplitude as the cantilever is scanned over the sample. A *tapping AFM* image is therefore produced by imaging the force of the intermittent contacts of the tip with the sample surface.

Although the peak forces applied during the contacting part of the oscillation can be much higher than typically used in contact mode, tapping mode generally lessens the damage done to the surface and the tip compared to the amount done in contact mode. This can be explained by the short duration of the applied force, and because the lateral forces between tip and sample are significantly lower in tapping mode over contact mode. Tapping mode imaging is gentle enough even for the visualization of supported lipid bilayers or adsorbed single polymer molecules (for instance, 0.4 nm thick chains of synthetic polyelectrolytes) under liquid medium. With proper scanning parameters, the conformation of single molecules can remain unchanged for hours, and even single molecular motors can be imaged while moving.

When operating in tapping mode, the phase of the cantilever's oscillation with respect to the driving signal can be recorded as well. This signal channel contains information about the energy dissipated by the cantilever in each oscillation cycle. Samples that contain regions of varying stiffness or with different adhesion properties can give a contrast in this channel that is not visible in the topographic image. Extracting the sample's material properties in a quantitative manner from phase images, however, is often not feasible.

Non-contact Mode

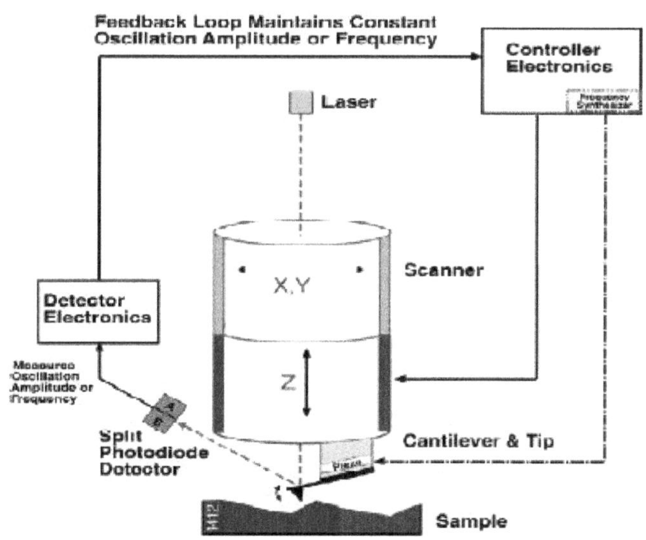

AFM – non-contact mode

In non-contact atomic force microscopy mode, the tip of the cantilever does not contact the sample surface. The cantilever is instead oscillated at either its resonant frequency (frequency modulation) or just above (amplitude modulation) where the amplitude of oscillation is typically a few nanometers (<10 nm) down to a few picometers. The van der Waals forces, which are strongest from 1 nm to 10 nm above the surface, or any other long-range force that extends above the surface acts to decrease the resonance frequency of the cantilever. This decrease in resonant frequency combined with the feedback loop system maintains a constant oscillation amplitude or frequency by adjusting the average tip-to-sample distance. Measuring the tip-to-sample distance at each (x,y) data point allows the scanning software to construct a topographic image of the sample surface.

Non-contact mode AFM does not suffer from tip or sample degradation effects that are sometimes observed after taking numerous scans with contact AFM. This makes non-contact AFM preferable to contact AFM for measuring soft samples, e.g. biological samples and organic thin film. In the case of rigid samples, contact and non-contact images may look the same. However, if a few monolayers of adsorbed fluid are lying on the surface of a rigid sample, the images may look quite different. An AFM operating in contact mode will penetrate the liquid layer to image the underlying surface, whereas in non-contact mode an AFM will oscillate above the adsorbed fluid layer to image both the liquid and surface.

Schemes for dynamic mode operation include frequency modulation where a phase-locked loop is used to track the cantilever's resonance frequency and the more common amplitude modulation with a servo loop in place to keep the cantilever excitation to a defined amplitude. In frequency modulation, changes in the oscillation frequency provide information about tip-sample interactions. Frequency can be measured with very high sensitivity and thus the frequency modulation mode allows for the use of very stiff cantilevers. Stiff cantilevers provide stability very close to the surface and, as a result, this technique was the first AFM technique to provide true atomic resolution in ultra-high vacuum conditions.

In amplitude modulation, changes in the oscillation amplitude or phase provide the feedback signal for imaging. In amplitude modulation, changes in the phase of oscillation can be used to discriminate between different types of materials on the surface. Amplitude modulation can be operated either in the non-contact or in the intermittent contact regime. In dynamic contact mode, the cantilever is oscillated such that the separation distance between the cantilever tip and the sample surface is modulated.

Amplitude modulation has also been used in the non-contact regime to image with atomic resolution by using very stiff cantilevers and small amplitudes in an ultra-high vacuum environment.

Topographic Image

Image formation is a plotting method that produces a color mapping through changing the x-y position of the tip while scanning and recording the measured variable, i.e. the intensity of control signal, to each x-y coordinate. The color mapping shows the measured value corresponding to each coordinate. The image expresses the intensity of a value as a hue. Usually, the correspondence between the intensity of a value and a hue is shown as a color scale in the explanatory notes accompanying the image.

What is the Topographic Image of Atomic-force Microscope?

Operation mode of Image forming of the AFM are generally classified into two groups from the viewpoint whether it uses z-Feedback loop (not shown) to maintain the tip-sample distance to keep signal intensity exported by the detector. The first one (using z-Feedback loop), said to be "constant XX mode" (XX is something which kept by z-Feedback loop).

Topographic Image Formation Mode is based on abovementioned "constant XX mode", z-Feedback loop controls the relative distance between the probe and the sample through outputting control signals to keep constant one of frequency, vibration and phase which typically corresponds to the motion of cantilever (for instance, voltage is applied to the Z-piezoelectric element and it moves the sample up and down towards the Z direction.

Topographic Image of FM-AFM

When the distance between the probe and the sample is brought to the range where atomic force may be detected, while a cantilever is excited in its natural eigen frequency (f_0), a phenomenon that the resonance frequency (f) of the cantilever shifts from the original resonance frequency (natural eigen frequency) of the cantilever. In other words, in the range where atomic force may be detected, the frequency shift ($df=f-f_0$) will be observed. So, when the distance between the probe and the sample is in the non-contact region, the frequency shift increases in negative direction as the distance between the probe and the sample gets smaller.

When the sample has concavity and convexity, the distance between the tip-apex and the sample varies in accordance with the concavity and convexity accompanied with a scan of the sample along x-y direction (without height regulation in z-direction). As a result, the frequency shift arises. The image in which the values of the frequency obtained by a raster scan along the x-y direction of the sample surface are plotted against the x-y coordination of each measurement point is called a constant-height image.

On the other hand, the df may be kept constant by moving the probe upward and downward in z-direction using a negative feedback (by using z-feedback loop) while the raster scan of the sample surface along the x-y direction. The image in which the amounts of the negative feedback (the moving distance of the probe upward and downward in z-direction) are plotted against the x-y coordination of each measurement point is a topographic image. In other words, the topographic image is a trace of the tip of the probe regulated so that the df is constant and it may also be considered to be a plot of a constant-height surface of the df.

Therefore, the topographic image of the AFM is not the exact surface morphology itself, but actually the image influenced by the bond-order between the probe and the sample, however, the topographic image of the AFM is considered to reflect the geographical shape of the surface more than the topographic image of a scanning tunnel microscope.

Force Spectroscopy

Another major application of AFM (besides imaging) is force spectroscopy, the direct measurement of tip-sample interaction forces as a function of the gap between the tip and sample (the result of this measurement is called a force-distance curve). For this method, the AFM tip is extended towards and retracted from the surface as the deflection of the cantilever is monitored as a function of piezoelectric displacement. These measurements have been used to measure nanoscale contacts, atomic bonding, Van der Waals forces, and Casimir forces, dissolution forces in liquids and single molecule stretching and rupture forces. Furthermore, AFM was used to measure, in an aqueous environment, the dispersion force due to polymer adsorbed on the substrate. Forces of the order of a few piconewtons can now be routinely measured with a vertical distance resolution of better than 0.1 nanometers. Force spectroscopy can be performed with either static or dynamic modes. In dynamic modes, information about the cantilever vibration is monitored in addition to the static deflection.

Problems with the technique include no direct measurement of the tip-sample separation and the common need for low-stiffness cantilevers, which tend to 'snap' to the surface. These problems are not insurmountable. An AFM that directly measures the tip-sample separation has been developed. The snap-in can be reduced by measuring in liquids or by using stiffer cantilevers, but in the latter case a more sensitive deflection sensor is needed. By applying a small dither to the tip, the stiffness (force gradient) of the bond can be measured as well.

Biological Applications and Other

Force spectroscopy is used in biophysics to measure the mechanical properties. of living material (such as tissue or cells). Another application was to measure the interaction forces between from one hand a material stuck on the tip of the cantilever, and from another hand the surface of particles either free or occupied by the same material. From the adhesion force distribution curve, a mean value of the forces has been derived. It allowed to make a cartography of the surface of the particles, covered or not by the material.

Identification of Individual Surface Atoms

The AFM can be used to image and manipulate atoms and structures on a variety of surfaces. The atom at the apex of the tip "senses" individual atoms on the underlying surface when it forms incipient chemical bonds with each atom. Because these chemical interactions subtly alter the tip's vibration frequency, they can be detected and mapped. This principle was used to distinguish between atoms of silicon, tin and lead on an alloy surface, by comparing these 'atomic fingerprints' to values obtained from large-scale density functional theory (DFT) simulations.

The trick is to first measure these forces precisely for each type of atom expected in the sample, and then to compare with forces given by DFT simulations. The team found that the tip interacted most strongly with silicon atoms, and interacted 24% and 41% less strongly with tin and lead atoms, respectively. Thus, each different type of atom can be identified in the matrix as the tip is moved across the surface.

Probe

An AFM probe has a sharp tip on the free-swinging end of a cantilever that is protruding from a holder. The dimensions of the cantilever are in the scale of micrometers. The radius of the tip is usually on the scale of a few nanometers to a few tens of nanometers. (Specialized probes exist with much larger end radii, for example probes for indentation of soft materials.) The cantilever holder, also called holder chip – often 1.6 mm by 3.4 mm in size – allows the operator to hold the AFM cantilever/probe assembly with tweezers and fit it into the corresponding holder clips on the scanning head of the atomic-force microscope.

This device is most commonly called an "AFM probe", but other names include "AFM tip" and "cantilever" (employing the name of a single part as the name of the whole device). An AFM probe is a particular type of SPM (scanning probe microscopy) probe.

AFM probes are manufactured with MEMS technology. Most AFM probes used are made from silicon (Si), but borosilicate glass and silicon nitride are also in use. AFM probes are considered consumables as they are often replaced when the tip apex becomes dull or contaminated or when the cantilever is broken. They can cost from a couple of tens of dollars up to hundreds of dollars per cantilever for the most specialized cantilever/probe combinations.

Just the tip is brought very close to the surface of the object under investigation, the cantilever is deflected by the interaction between the tip and the surface, which is what the AFM is designed to measure. A spatial map of the interaction can be made by measuring the deflection at many points on a 2D surface.

Several types of interaction can be detected. Depending on the interaction under investigation, the surface of the tip of the AFM probe needs to be modified with a coating. Among the coatings used are gold – for covalent bonding of biological molecules and the detection of their interaction with a surface, diamond for increased wear resistance and magnetic coatings for detecting the magnetic properties of the investigated surface. Another solution exists to achieve high resolution magnetic imaging : having the probe equip with a microSQUID. The AFM tips is fabricated using silicon micro machining and the precise positioning of the microSQUID loop is done by electron beam lithography.

The surface of the cantilevers can also be modified. These coatings are mostly applied in order to increase the reflectance of the cantilever and to improve the deflection signal.

AFM Cantilever-deflection Measurement

Beam-deflection Measurement

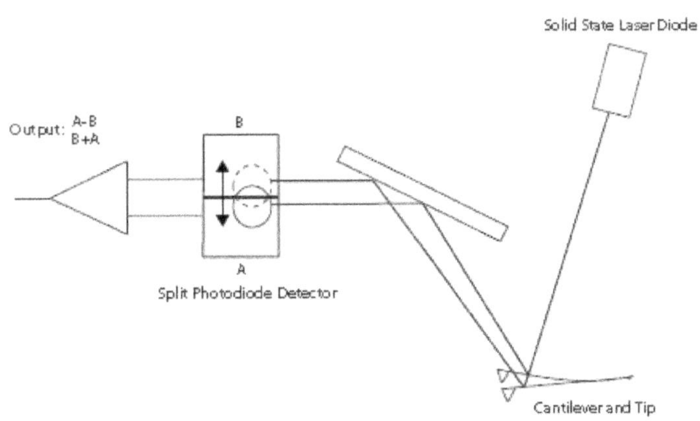

AFM beam-deflection detection

The most common method for cantilever-deflection measurements is the beam-deflection method. In this method, laser light from a solid-state diode is reflected off the back of the cantilever and collected by a position-sensitive detector (PSD) consisting of two closely spaced photodiodes, whose output signal is collected by a differential amplifier. Angular displacement of the cantilever results in one photodiode collecting more light than the other photodiode, producing an output signal (the difference between the photodiode signals normalized by their sum), which is proportional to the deflection of the cantilever. The sensitivity of the beam-deflection method is very high, a noise floor on the order of 10 fm $Hz^{-1/2}$ can be obtained routinely in a well-designed system. Although this method is sometimes called the 'optical lever' method, the signal is not amplified if the beam path is made longer. A longer beam path increases the motion of the reflected spot on the photodiodes, but also widens the spot by the same amount due to diffraction, so that the same amount of optical power is moved from one photodiode to the other. The 'optical leverage' (output signal of the detector divided by deflection of the cantilever) is inversely proportional to the numerical aperture of the beam focusing optics, as long as the focused laser spot is small enough to fall completely on the cantilever. It is also inversely proportional to the length of the cantilever.

The relative popularity of the beam-deflection method can be explained by its high sensitivity and simple operation, and by the fact that cantilevers do not require electrical contacts or other special treatments, and can therefore be fabricated relatively cheaply with sharp integrated tips.

Other Deflection-measurement Methods

Many other methods for beam-deflection measurements exist.

- *Piezoelectric detection* – Cantilevers made from quartz (such as the qPlus configuration), or other piezoelectric materials can directly detect deflection as an electrical signal. Cantilever oscillations down to 10pm have been detected with this method.
- *Laser Doppler vibrometry* – A laser Doppler vibrometer can be used to produce very accurate deflection measurements for an oscillating cantilever (thus is only used in non-contact mode). This method is expensive and is only used by relatively few groups.
- *STM* — The first atomic microscope used an STM complete with its own feedback mechanism to measure deflection. This method is very difficult to implement, and is slow to react to deflection changes compared to modern methods.
- *Optical interferometry* – Optical interferometry can be used to measure cantilever deflection. Due to the nanometre scale deflections measured in AFM, the interferometer is running in the sub-fringe regime, thus, any drift in laser power or wavelength has strong effects on the measurement. For these reasons optical interferometer measurements must be done with great care (for example using index matching fluids between optical fibre junctions), with very stable lasers. For these reasons optical interferometry is rarely used.
- *Capacitive detection* – Metal coated cantilevers can form a capacitor with another contact located behind the cantilever. Deflection changes the distance between the contacts and can be measured as a change in capacitance.

- *Piezoresistive detection* – Cantilevers can be fabricated with piezoresistive elements that act as a strain gauge. Using a Wheatstone bridge, strain in the AFM cantilever due to deflection can be measured. This is not commonly used in vacuum applications, as the piezoresistive detection dissipates energy from the system affecting Q of the resonance.

Piezoelectric Scanners

AFM scanners are made from piezoelectric material, which expands and contracts proportionally to an applied voltage. Whether they elongate or contract depends upon the polarity of the voltage applied. Traditionally the tip or sample is mounted on a 'tripod' of three piezo crystals, with each responsible for scanning in the *x,y* and *z* directions. In 1986, the same year as the AFM was invented, a new piezoelectric scanner, the tube scanner, was developed for use in STM. Later tube scanners were incorporated into AFMs. The tube scanner can move the sample in the *x*, *y*, and *z* directions using a single tube piezo with a single interior contact and four external contacts. An advantage of the tube scanner compared to the original tripod design, is better vibrational isolation, resulting from the higher resonant frequency of the single element construction, in combination with a low resonant frequency isolation stage. A disadvantage is that the *x-y* motion can cause unwanted *z* motion resulting in distortion. Another popular design for AFM scanners is the flexure stage, which uses separate piezos for each axis, and couples them through a flexure mechanism.

Scanners are characterized by their sensitivity, which is the ratio of piezo movement to piezo voltage, i.e., by how much the piezo material extends or contracts per applied volt. Because of differences in material or size, the sensitivity varies from scanner to scanner. Sensitivity varies non-linearly with respect to scan size. Piezo scanners exhibit more sensitivity at the end than at the beginning of a scan. This causes the forward and reverse scans to behave differently and display hysteresis between the two scan directions. This can be corrected by applying a non-linear voltage to the piezo electrodes to cause linear scanner movement and calibrating the scanner accordingly. One disadvantage of this approach is that it requires re-calibration because the precise non-linear voltage needed to correct non-linear movement will change as the piezo ages. This problem can be circumvented by adding a linear sensor to the sample stage or piezo stage to detect the true movement of the piezo. Deviations from ideal movement can be detected by the sensor and corrections applied to the piezo drive signal to correct for non-linear piezo movement. This design is known as a 'closed loop' AFM. Non-sensored piezo AFMs are referred to as 'open loop' AFMs.

The sensitivity of piezoelectric materials decreases exponentially with time. This causes most of the change in sensitivity to occur in the initial stages of the scanner's life. Piezoelectric scanners are run for approximately 48 hours before they are shipped from the factory so that they are past the point where they may have large changes in sensitivity. As the scanner ages, the sensitivity will change less with time and the scanner would seldom require recalibration, though various manufacturer manuals recommend monthly to semi-monthly calibration of open loop AFMs.

Advantages and Disadvantages

Just like any other tool, an AFM's usefulness has limitations. When determining whether or not analyzing a sample with an AFM is appropriate, there are various advantages and disadvantages that must be considered.

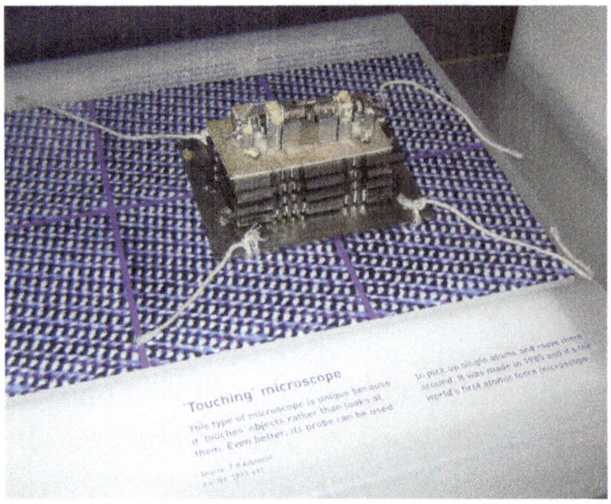

The first atomic-force microscope

Advantages

AFM has several advantages over the scanning electron microscope (SEM). Unlike the electron microscope, which provides a two-dimensional projection or a two-dimensional image of a sample, the AFM provides a three-dimensional surface profile. In addition, samples viewed by AFM do not require any special treatments (such as metal/carbon coatings) that would irreversibly change or damage the sample, and does not typically suffer from charging artifacts in the final image. While an electron microscope needs an expensive vacuum environment for proper operation, most AFM modes can work perfectly well in ambient air or even a liquid environment. This makes it possible to study biological macromolecules and even living organisms. In principle, AFM can provide higher resolution than SEM. It has been shown to give true atomic resolution in ultra-high vacuum (UHV) and, more recently, in liquid environments. High resolution AFM is comparable in resolution to scanning tunneling microscopy and transmission electron microscopy. AFM can also be combined with a variety of optical microscopy techniques such as fluorescent microscopy, further expanding its applicability. Combined AFM-optical instruments have been applied primarily in the biological sciences but have also found a niche in some materials applications, especially those involving photovoltaics research.

Disadvantages

A disadvantage of AFM compared with the scanning electron microscope (SEM) is the single scan image size. In one pass, the SEM can image an area on the order of square millimeters with a depth of field on the order of millimeters, whereas the AFM can only image a maximum scanning area of about 150×150 micrometers and a maximum height on the order of 10-20 micrometers. One method of improving the scanned area size for AFM is by using parallel probes in a fashion similar to that of millipede data storage.

The scanning speed of an AFM is also a limitation. Traditionally, an AFM cannot scan images as fast as an SEM, requiring several minutes for a typical scan, while an SEM is capable of scanning at near real-time, although at relatively low quality. The relatively slow rate of scanning during AFM imaging often leads to thermal drift in the image making the AFM less suited for measuring accurate distances between topographical features on the image. However, several fast-acting designs were suggested to increase microscope scanning productivity including what is being termed videoAFM (reasonable quality images are being obtained with videoAFM at video rate: faster than the average SEM). To eliminate image distortions induced by thermal drift, several methods have been introduced.

AFM images can also be affected by nonlinearity, hysteresis, and creep of the piezoelectric material and cross-talk between the x, y, z axes that may require software enhancement and filtering. Such filtering could "flatten" out real topographical features. However, newer AFMs utilize real-time correction software (for example, feature-oriented scanning) or closed-loop scanners, which practically eliminate these problems. Some AFMs also use separated orthogonal scanners (as opposed to a single tube), which also serve to eliminate part of the cross-talk problems.

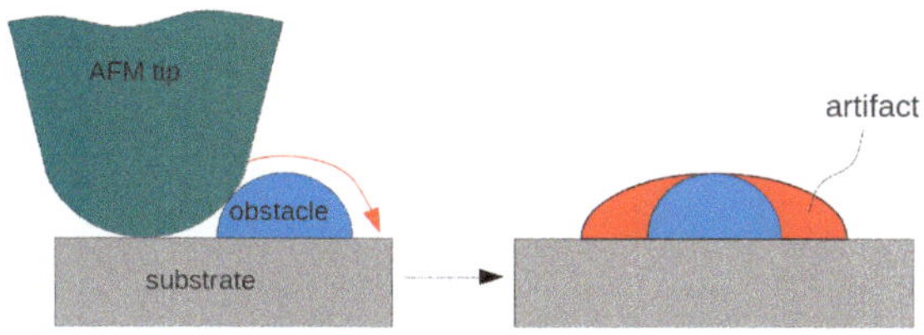

Showing an AFM artifact arising from a tip with a high radius of curvature with respect to the feature that is to be visualized.

As with any other imaging technique, there is the possibility of image artifacts, which could be induced by an unsuitable tip, a poor operating environment, or even by the sample itself, as depicted on the right. These image artifacts are unavoidable; however, their occurrence and effect on results can be reduced through various methods. Artifacts resulting from a too-coarse tip can be caused for example by inappropriate handling or de facto collisions with the sample by either scanning too fast or having an unreasonably rough surface, causing actual wearing of the tip.

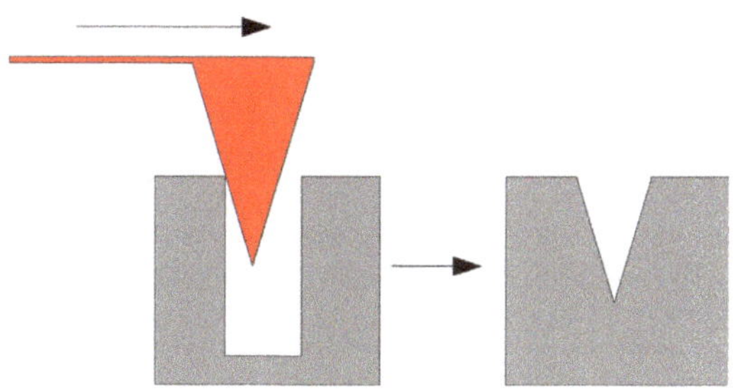

AFM artifact, steep sample topography

Due to the nature of AFM probes, they cannot normally measure steep walls or overhangs. Specially made cantilevers and AFMs can be used to modulate the probe sideways as well as up and down (as with dynamic contact and non-contact modes) to measure sidewalls, at the cost of more expensive cantilevers, lower lateral resolution and additional artifacts.

Other Applications in Various Fields of Study

The latest efforts in integrating nanotechnology and biological research have been successful and show much promise for the future. Since nanoparticles are a potential vehicle of drug delivery, the biological responses of cells to these nanoparticles are continuously being explored to optimize their efficacy and how their design could be improved. Pyrgiotakis et al. were able to study the interaction between CeO_2 and Fe_2O_3 engineered nanoparticles and cells by attaching the engineered nanoparticles to the AFM tip. Beyond the interactions with external synthetic materials, cells have been imaged with X-ray crystallography and there has been much curiosity about their behavior *in vivo*. Studies have taken advantage of AFM to obtain further information on the behavior of live cells in biological media. Real-time atomic force spectroscopy (or nanoscopy) and dynamic atomic force spectroscopy have been used to study live cells and membrane proteins and their dynamic behavior at high resolution, on the nanoscale. Imaging and obtaining information on the topography and the properties of the cells has also given insight into chemical processes and mechanisms that occur through cell-cell interaction and interactions with other signaling molecules (ex. ligands). Evans and Calderwood used single cell force microscopy to study cell adhesion forces, bond kinetics/dynamic bond strength and its role in chemical processes such as cell signaling. Scheuring, Lévy, and Rigaud reviewed studies in which AFM to explore the crystal structure of membrane proteins of photosynthetic bacteria. Alsteen et al. have used AFM-based nanoscopy to perform a real-time analysis of the interaction between live mycobacteria and antimycobacterial drugs (specifically isoniazid, ethionamide, ethambutol, and streptomycine), which serves as an example of the more in-depth analysis of pathogen-drug interactions that can be done through AFM.

References

- Smith KCA, Oatley, CW (1955). "The scanning electron microscope and its fields of application". British Journal of Applied Physics. 6 (11): 391–399. Bibcode:1955BJAP....6..391S. doi:10.1088/0508-3443/6/11/304
- Stokes, Debbie J. (2008). Principles and Practice of Variable Pressure Environmental Scanning Electron Microscopy (VP-ESEM). Chichester: John Wiley & Sons. ISBN 978-0470758748
- "Next Monday, Digital Surf to Launch Revolutionary SEM Image Colorization". AZO Materials. Retrieved 23 January 2016
- Suzuki, E. (2002). "High-resolution scanning electron microscopy of immunogold-labelled cells by the use of thin plasma coating of osmium". Journal of Microscopy. 208 (3): 153–157. doi:10.1046/j.1365-2818.2002.01082.x
- Goldstein, G. I.; Newbury, D. E.; Echlin, P.; Joy, D. C.; Fiori, C.; Lifshin, E. (1981). Scanning electron microscopy and x-ray microanalysis. New York: Plenum Press. ISBN 0-306-40768-X
- Geisse, Nicholas A. (July–August 2009). "AFM and Combined Optical Techniques". Materials Today. 12 (7–8): 40–45. doi:10.1016/S1369-7021(09)70201-9. Retrieved 4 November 2011
- Karnovsky, M. J. (1965). "A formaldehyde-glutaraldehyde fixative of high osmolality for use in electron microscopy" (PDF). Journal of Cell Biology. 27 (2): 137A. JSTOR 1604673
- Jeffree, C. E.; Read, N. D. (1991). "Ambient- and Low-temperature scanning electron microscopy". In Hall, J. L.; Hawes, C. R. Electron Microscopy of Plant Cells. London: Academic Press. pp. 313–413. ISBN 0-12-318880-6
- Hindmarsh, J. P.; Russell, A. B.; Chen, X. D. (2007). "Fundamentals of the spray freezing of foods—microstructure of frozen droplets". Journal of Food Engineering. 78 (1): 136–150. doi:10.1016/j.jfoodeng.2005.09.011
- Chandler, Douglas E.; Roberson, Robert W. (2009). Bioimaging : current concepts in light and electron microscopy. Sudbury, Mass.: Jones and Bartlett Publishers. ISBN 9780763738747
- Hasselbach, K.; Ladam, C. (2008). "High resolution magnetic imaging : MicroSQUID Force Microscopy". Journal of Physics: Conference Series. 97: 012330. Bibcode:2008JPhCS..97a2330H. doi:10.1088/1742-6596/97/1/012330
- Göddenhenrich, T. "Force microscope with capacitive displacement detection". Journal of Vacuum Science and Technology A. 8 (1): 383. doi:10.1116/1.576401

- Oura, K.; Lifshits, V.G.; Saranin, A.A.; Zotov, A.V.; et al. (2003). Surface Science: An Introduction. Springer-Verlag. ISBN 3-540-00545-5
- Jong, Wim H De; Borm, Paul JA (June 2008). "Drug Delivery and Nanoparticles: Applications and Hazards". Dovepress International Journal of Nanomedicine: 133–149
- R. V. Lapshin (2011). "Feature-oriented scanning probe microscopy". In H. S. Nalwa. Encyclopedia of Nanoscience and Nanotechnology (PDF). 14. USA: American Scientific Publishers. pp. 105–115. ISBN 1-58883-163-9
- Perkins, Thomas. "Atomic force microscopy measures properties of proteins and protein folding". SPIE Newsroom. Retrieved 4 March 2016

4

Diffraction Techniques for Materials Characterization

X-rays are electromagnetic radiation with wavelength ranging from 0.01 to 10 nanometers. X-rays are used for structural analysis of materials. It is known to interact with matter through photoelectric absorption, Compton scattering, and Rayleigh scattering. The chapter also explores the concepts of Bragg's law, powder diffraction and crystallite among few others. Tools are an important component of any field of study. The following chapter elucidates the various tools that are related to material characterization.

X-ray Diffraction

In 1885, Wilhelm Conrad Rontgen, a German physicist, discovered X-rays.

Wilhelm Conrad Rontgen

These are electromagnetic radiations with a wavelength ranging from 0.01 to 100 Å, which corresponds to the frequency in the range of 30 PHz - 300 EHz.

X-radiation (composed of X-rays) is a form of electromagnetic radiation. Most X-rays have a wavelength ranging from 0.01 to 10 nanometers, corresponding to frequencies in the range 30 petahertz to 30 exahertz (3×10^{16} Hz to 3×10^{19} Hz) and energies in the range 100 eV to 100 keV. X-ray wavelengths are shorter than those of UV rays and typically longer than those of gamma rays. In many languages, X-radiation is referred to with terms meaning Röntgen radiation, after the German/Dutch scientist Wilhelm Röntgen, who usually is credited as its discoverer, and who had named it *X-radiation* to signify an unknown type of radiation. Spelling of *X-ray(s)* in the English language includes the variants *x-ray(s)*, *xray(s)*, and *X ray(s)*.

X-rays with high photon energies (above 5–10 keV, below 0.2–0.1 nm wavelength) are called *hard X-rays*, while those with lower energy are called *soft X-rays*. Due to their penetrating ability, hard X-rays are widely used to image the inside of objects, e.g., in medical radiography and airport security. The term *X-ray* is metonymically used to refer to a radiographic image produced using this method, in addition to the method itself. Since the wavelengths of hard X-rays are similar to the size of atoms they are also useful for determining crystal structures by X-ray crystallography. By contrast, soft X-rays are easily absorbed in air; the attenuation length of 600 eV (~2 nm) X-rays in water is less than 1 micrometer.

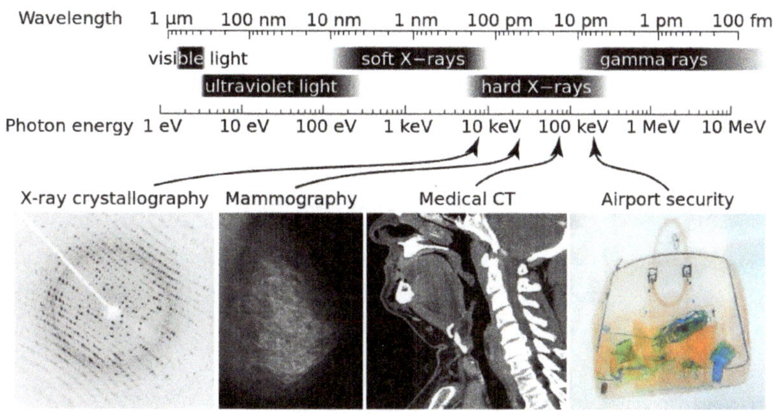

X-rays are part of the electromagnetic spectrum, with wavelengths shorter than visible light. Different applications use different parts of the X-ray spectrum.

There is no consensus for a definition distinguishing between X-rays and gamma rays. One common practice is to distinguish between the two types of radiation based on their source: X-rays are emitted by electrons, while gamma rays are emitted by the atomic nucleus. This definition has several problems: other processes also can generate these high-energy photons, or sometimes the method of generation is not known. One common alternative is to distinguish X- and gamma radiation on the basis of wavelength (or, equivalently, frequency or photon energy), with radiation shorter than some arbitrary wavelength, such as 10^{-11} m (0.1 Å), defined as gamma radiation. This criterion assigns a photon to an unambiguous category, but is only possible if wavelength is known. (Some measurement techniques do not distinguish between detected wavelengths.) However, these two definitions often coincide since the electromagnetic radiation emitted by X-ray tubes generally has a longer wavelength and lower photon energy than the radiation emitted by radioactive nuclei. Occasionally, one term or the other is used in specific contexts due to historical precedent, based on measurement (detection) technique, or based on their intended use rather than their wavelength or source. Thus, gamma-rays generated for medical and industrial uses, for example radiotherapy, in the ranges of 6–20 MeV, can in this context also be referred to as X-rays.

Properties

X-ray photons carry enough energy to ionize atoms and disrupt molecular bonds. This makes it a type of ionizing radiation, and therefore harmful to living tissue. A very high radiation dose over a short period of time causes radiation sickness, while lower doses can give an increased risk of radiation-induced cancer. In medical imaging this increased cancer risk is generally greatly outweighed by the benefits of the examination. The ionizing capability of X-rays can be utilized in cancer treatment to kill malignant cells using radiation therapy. It is also used for material characterization using X-ray spectroscopy.

Ionizing radiation hazard symbol

Hard X-rays can traverse relatively thick objects without being much absorbed or scattered. For this reason, X-rays are widely used to image the inside of visually opaque objects. The most often seen applications are in medical radiography and airport security scanners, but similar techniques are also important in industry (e.g. industrial radiography and industrial CT scanning) and research (e.g. small animal CT). The penetration depth varies with several orders of magnitude over the X-ray spectrum. This allows the photon energy to be adjusted for the application so as to give sufficient transmission through the object and at the same time good contrast in the image.

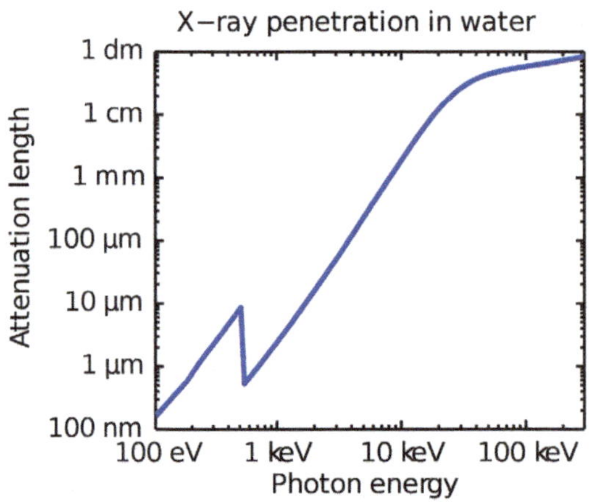

Attenuation length of X-rays in water showing the oxygen absorption edge at 540 eV, the energy^{-3} dependence of photoabsorption, as well as a leveling off at higher photon energies due to Compton scattering. The attenuation length is about four orders of magnitude longer for hard X-rays (right half) compared to soft X-rays (left half).

X-rays have much shorter wavelength than visible light, which makes it possible to probe structures much smaller than what can be seen using a normal microscope. This can be used in X-ray microscopy to acquire high resolution images, but also in X-ray crystallography to determine the positions of atoms in crystals.

Interaction with Matter

X-rays interact with matter in three main ways, through photoabsorption, Compton scattering, and Rayleigh scattering. The strength of these interactions depends on the energy of the X-rays and the elemental composition of the material, but not much on chemical properties, since the X-ray photon energy is much higher than chemical binding energies. Photoabsorption or photoelectric absorption is the dominant interaction mechanism in the soft X-ray regime and for the lower hard X-ray energies. At higher energies, Compton scattering dominates.

Photoelectric Absorption

The probability of a photoelectric absorption per unit mass is approximately proportional to Z^3/E^3, where Z is the atomic number and E is the energy of the incident photon. This rule is not valid close to inner shell electron binding energies where there are abrupt changes in interaction probability, so called absorption edges. However, the general trend of high absorption coefficients and thus short penetration depths for low photon energies and high atomic numbers is very strong. For soft tissue photoabsorption dominates up to about 26 keV photon energy where Compton scattering takes over. For higher atomic number substances this limit is higher. The high amount of calcium (Z=20) in bones together with their high density is what makes them show up so clearly on medical radiographs.

A photoabsorbed photon transfers all its energy to the electron with which it interacts, thus ionizing the atom to which the electron was bound and producing a photoelectron that is likely to ionize more atoms in its path. An outer electron will fill the vacant electron position and produce either a characteristic photon or an Auger electron. These effects can be used for elemental detection through X-ray spectroscopy or Auger electron spectroscopy.

Compton Scattering

Compton scattering is the predominant interaction between X-rays and soft tissue in medical imaging. Compton scattering is an inelastic scattering of the X-ray photon by an outer shell electron. Part of the energy of the photon is transferred to the scattering electron, thereby ionizing the atom and increasing the wavelength of the X-ray. The scattered photon can go in any direction, but a direction similar to the original direction is more likely, especially for high-energy X-rays. The probability for different scattering angles are described by the Klein–Nishina formula. The transferred energy can be directly obtained from the scattering angle from the conservation of energy and momentum.

Rayleigh Scattering

Rayleigh scattering is the dominant elastic scattering mechanism in the X-ray regime. Inelastic forward scattering gives rise to the refractive index, which for X-rays is only slightly below 1.

Production

Whenever charged particles (electrons or ions) of sufficient energy hit a material, x-rays are produced.

Production by Electrons

Characteristic X-ray emission lines for some common anode materials.

Anode material	Atomic number	Photon energy [keV]		Wavelength [nm]	
		$K_{\alpha 1}$	$K_{\beta 1}$	$K_{\alpha 1}$	$K_{\beta 1}$
W	74	59.3	67.2	0.0209	0.0184
Mo	42	17.5	19.6	0.0709	0.0632
Cu	29	8.05	8.91	0.154	0.139

Ag	47	22.2	24.9	0.0559	0.0497
Ga	31	9.25	10.26	0.134	0.121
In	49	24.2	27.3	0.0512	0.455

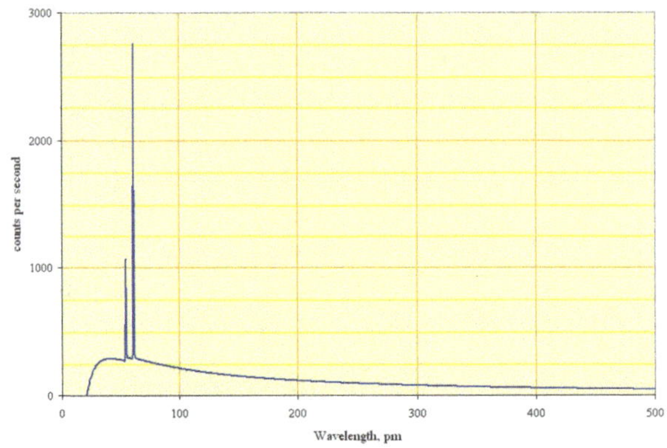

Spectrum of the X-rays emitted by an X-ray tube with a rhodium target, operated at 60 kV. The smooth, continuous curve is due to *bremsstrahlung*, and the spikes are characteristic K lines for rhodium atoms.

X-rays can be generated by an X-ray tube, a vacuum tube that uses a high voltage to accelerate the electrons released by a hot cathode to a high velocity. The high velocity electrons collide with a metal target, the anode, creating the X-rays. In medical X-ray tubes the target is usually tungsten or a more crack-resistant alloy of rhenium (5%) and tungsten (95%), but sometimes molybdenum for more specialized applications, such as when softer X-rays are needed as in mammography. In crystallography, a copper target is most common, with cobalt often being used when fluorescence from iron content in the sample might otherwise present a problem.

The maximum energy of the produced X-ray photon is limited by the energy of the incident electron, which is equal to the voltage on the tube times the electron charge, so an 80 kV tube cannot create X-rays with an energy greater than 80 keV. When the electrons hit the target, X-rays are created by two different atomic processes:

1. *Characteristic X-ray emission* (X-ray fluorescence): If the electron has enough energy it can knock an orbital electron out of the inner electron shell of a metal atom, and as a result electrons from higher energy levels then fill up the vacancy and X-ray photons are emitted. This process produces an emission spectrum of X-rays at a few discrete frequencies, sometimes referred to as the spectral lines. The spectral lines generated depend on the target (anode) element used and thus are called characteristic lines. Usually these are transitions from upper shells into K shell (called K lines), into L shell (called L lines) and so on.

2. *Bremsstrahlung*: This is radiation given off by the electrons as they are scattered by the strong electric field near the high-*Z* (proton number) nuclei. These X-rays have a continuous spectrum. The intensity of the X-rays increases linearly with decreasing frequency, from zero at the energy of the incident electrons, the voltage on the X-ray tube.

So the resulting output of a tube consists of a continuous bremsstrahlung spectrum falling off to zero at the tube voltage, plus several spikes at the characteristic lines. The voltages used in diagnostic X-ray tubes range from roughly 20 kV to 150 kV and thus the highest energies of the X-ray photons range from roughly 20 keV to 150 keV.

Both of these X-ray production processes are inefficient, with a production efficiency of only about one percent, and thus most of the electric power consumed by the tube is released as waste heat. When producing a usable flux of X-rays, the X-ray tube must be designed to dissipate the excess heat.

Short nanosecond bursts of X-rays peaking at 15-keV in energy may be reliably produced by peeling pressure-sensitive adhesive tape from its backing in a moderate vacuum. This is likely to be the result of recombination of electrical charges produced by triboelectric charging. The intensity of X-ray triboluminescence is sufficient for it to be used as a source for X-ray imaging.

A specialized source of X-rays which is becoming widely used in research is synchrotron radiation, which is generated by particle accelerators. Its unique features are X-ray outputs many orders of magnitude greater than those of X-ray tubes, wide X-ray spectra, excellent collimation, and linear polarization.

Production by Fast Positive Ions

X-rays can also be produced by fast protons or other positive ions. The proton-induced X-ray emission or particle-induced X-ray emission is widely used as an analytical procedure. For high energies, the production cross section is proportional to $Z_1^2 Z_2^{-4}$, where Z_1 refers to the atomic number of the ion, Z_2 to that of the target atom.

Detectors

X-ray detectors vary in shape and function depending on their purpose. Imaging detectors such as those used for radiography were originally based on photographic plates and later photographic film, but are now mostly replaced by various digital detector types such as image plates and flat panel detectors. For radiation protection direct exposure hazard is often evaluated using ionization chambers, while dosimeters are used to measure the radiation dose a person has been exposed to. X-ray spectra can be measured either by energy dispersive or wavelength dispersive spectrometers.

Medical Uses

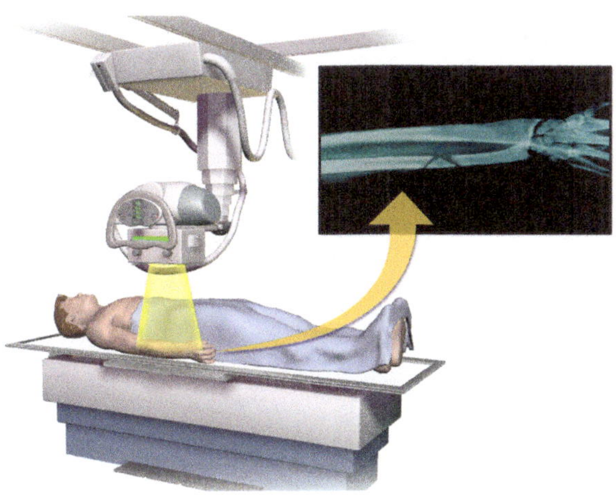

X-ray.

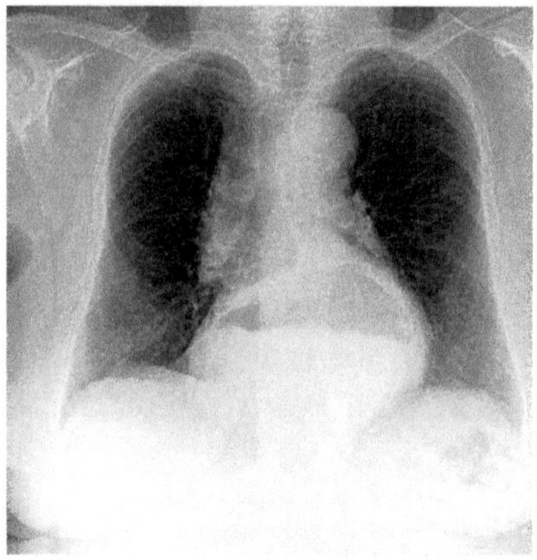

A chest radiograph of a female, demonstrating a hiatus hernia

Since Röntgen's discovery that X-rays can identify bone structures, X-rays have been used for medical imaging. The first medical use was less than a month after his paper on the subject. Up until 2010, 5 billion medical imaging studies have been conducted worldwide. Radiation exposure from medical imaging in 2006 made up about 50% of total ionizing radiation exposure in the United States.

Projectional Radiographs

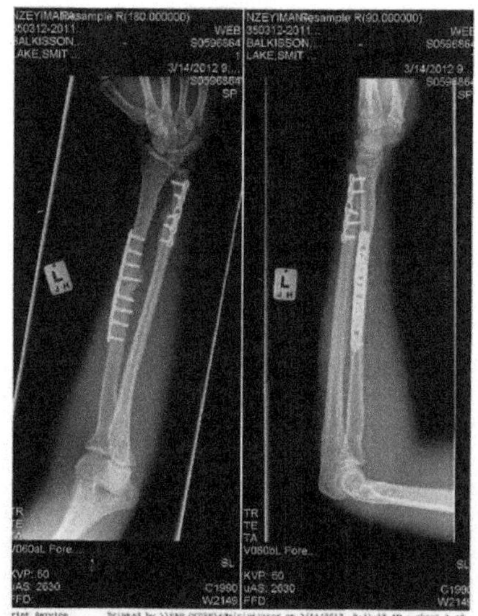

An arm radiograph, demonstrating broken ulna and radius with implanted internal fixation.

Projectional radiography is the practice of producing two-dimensional images using x-ray radiation. Bones contain much calcium, which due to its relatively high atomic number absorbs x-rays efficiently. This reduces the amount of X-rays reaching the detector in the shadow of the bones, making them clearly visible on the radiograph. The lungs and trapped gas also show up clearly because of lower absorption compared to tissue, while differences between tissue types are harder to see.

Projectional radiographs are useful in the detection of pathology of the skeletal system as well as for detecting some disease processes in soft tissue. Some notable examples are the very common chest X-ray, which can be used to identify lung diseases such as pneumonia, lung cancer, or pulmonary edema, and the abdominal x-ray, which can detect bowel (or intestinal) obstruction, free air (from visceral perforations) and free fluid (in ascites). X-rays may also be used to detect pathology such as gallstones (which are rarely radiopaque) or kidney stones which are often (but not always) visible. Traditional plain X-rays are less useful in the imaging of soft tissues such as the brain or muscle.

Dental radiography is commonly used in the diagnoses of common oral problems, such as cavities.

In medical diagnostic applications, the low energy (soft) X-rays are unwanted, since they are totally absorbed by the body, increasing the radiation dose without contributing to the image. Hence, a thin metal sheet, often of aluminium, called an X-ray filter, is usually placed over the window of the X-ray tube, absorbing the low energy part in the spectrum. This is called *hardening* the beam since it shifts the center of the spectrum towards higher energy (or harder) x-rays.

To generate an image of the cardiovascular system, including the arteries and veins (angiography) an initial image is taken of the anatomical region of interest. A second image is then taken of the same region after an iodinated contrast agent has been injected into the blood vessels within this area. These two images are then digitally subtracted, leaving an image of only the iodinated contrast outlining the blood vessels. The radiologist or surgeon then compares the image obtained to normal anatomical images to determine whether there is any damage or blockage of the vessel.

Computed Tomography

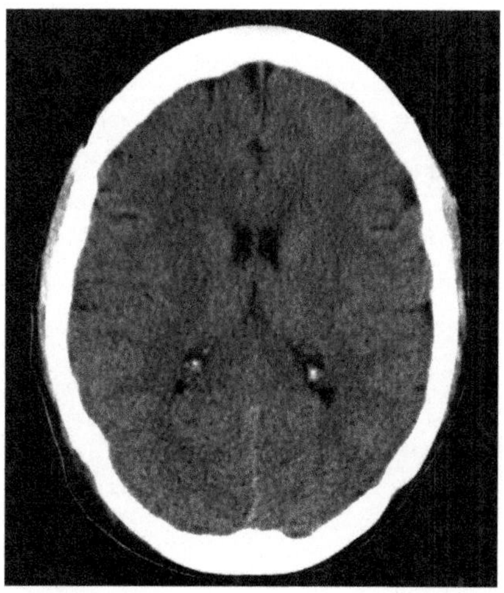

Head CT scan (transverse plane) slice -- a modern application of medical radiography

Computed tomography (CT scanning) is a medical imaging modality where tomographic images or slices of specific areas of the body are obtained from a large series of two-dimensional X-ray images taken in different directions. These cross-sectional images can be combined into a three-dimensional image of the inside of the body and used for diagnostic and therapeutic purposes in various medical disciplines.

Fluoroscopy

Fluoroscopy is an imaging technique commonly used by physicians or radiation therapists to obtain real-time moving images of the internal structures of a patient through the use of a fluoroscope. In its simplest form, a fluoroscope consists of an X-ray source and a fluorescent screen, between which a patient is placed. However, modern fluoroscopes couple the screen to an X-ray image intensifier and CCD video camera allowing the images to be recorded and played on a monitor. This method may use a contrast material. Examples include cardiac catheterization (to examine for coronary artery blockages) and barium swallow (to examine for esophageal disorders).

Radiotherapy

The use of X-rays as a treatment is known as radiation therapy and is largely used for the management (including palliation) of cancer; it requires higher radiation doses than those received for imaging alone. X-rays beams are used for treating skin cancers using lower energy x-ray beams while higher energy beams are used for treating cancers within the body such as brain, lung, prostate, and breast.

Adverse Effects

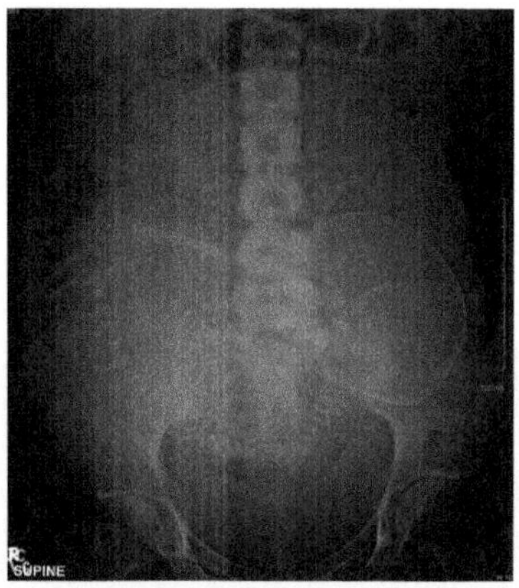

Abdominal radiograph of a pregnant woman, a procedure that should be performed only after proper assessment of benefit *versus* risk

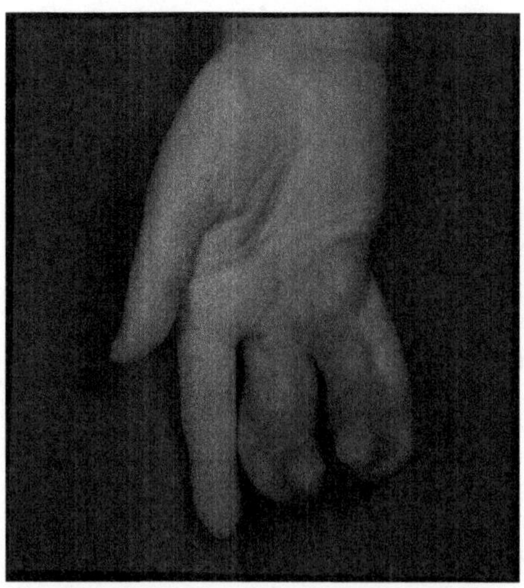

Deformity of hand due to an X-ray burn. These burns are accidents. X-rays were not shielded when they were first discovered and used, and people received radiation burns.

Diagnostic X-rays (primarily from CT scans due to the large dose used) increase the risk of developmental problems and cancer in those exposed. X-rays are classified as a carcinogen by both the World Health Organization's International Agency for Research on Cancer and the U.S. government. It is estimated that 0.4% of current cancers in the United States are due to computed tomography (CT scans) performed in the past and that this may increase to as high as 1.5-2% with 2007 rates of CT usage.

Experimental and epidemiological data currently do not support the proposition that there is a threshold dose of radiation below which there is no increased risk of cancer. However, this is under increasing doubt. It is estimated that the additional radiation will increase a person's cumulative risk of getting cancer by age 75 by 0.6–1.8%. The amount of absorbed radiation depends upon the type of X-ray test and the body part involved. CT and fluoroscopy entail higher doses of radiation than do plain X-rays.

To place the increased risk in perspective, a plain chest X-ray will expose a person to the same amount from background radiation that people are exposed to (depending upon location) every day over 10 days, while exposure from a dental X-ray is approximately equivalent to 1 day of environmental background radiation. Each such X-ray would add less than 1 per 1,000,000 to the lifetime cancer risk. An abdominal or chest CT would be the equivalent to 2–3 years of background radiation to the whole body, or 4–5 years to the abdomen or chest, increasing the lifetime cancer risk between 1 per 1,000 to 1 per 10,000. This is compared to the roughly 40% chance of a US citizen developing cancer during their lifetime. For instance, the effective dose to the torso from a CT scan of the chest is about 5 mSv, and the absorbed dose is about 14 mGy. A head CT scan (1.5mSv, 64mGy) that is performed once with and once without contrast agent, would be equivalent to 40 years of background radiation to the head. Accurate estimation of effective doses due to CT is difficult with the estimation uncertainty range of about ±19% to ±32% for adult head scans depending upon the method used.

The risk of radiation is greater to unborn babies, so in pregnant patients, the benefits of the investigation (X-ray) should be balanced with the potential hazards to the unborn fetus. In the US, there are an estimated 62 million CT scans performed annually, including more than 4 million on children. Avoiding unnecessary X-rays (especially CT scans) reduces radiation dose and any associated cancer risk.

Medical X-rays are a significant source of man-made radiation exposure. In 1987, they accounted for 58% of exposure from man-made sources in the United States. Since man-made sources accounted for only 18% of the total radiation exposure, most of which came from natural sources (82%), medical X-rays only accounted for 10% of *total* American radiation exposure; medical procedures as a whole (including nuclear medicine) accounted for 14% of total radiation exposure. By 2006, however, medical procedures in the United States were contributing much more ionizing radiation than was the case in the early 1980s. In 2006, medical exposure constituted nearly half of the total radiation exposure of the U.S. population from all sources. The increase is traceable to the growth in the use of medical imaging procedures, in particular computed tomography (CT), and to the growth in the use of nuclear medicine.

Dosage due to dental X-rays varies significantly depending on the procedure and the technology (film or digital). Depending on the procedure and the technology, a single dental X-ray of a human results in an exposure of 0.5 to 4 mrem. A full mouth series may therefore result in an exposure of up to 6 (digital) to 18 (film) mrem, for a yearly average of up to 40 mrem.

Other Uses

Other notable uses of X-rays include

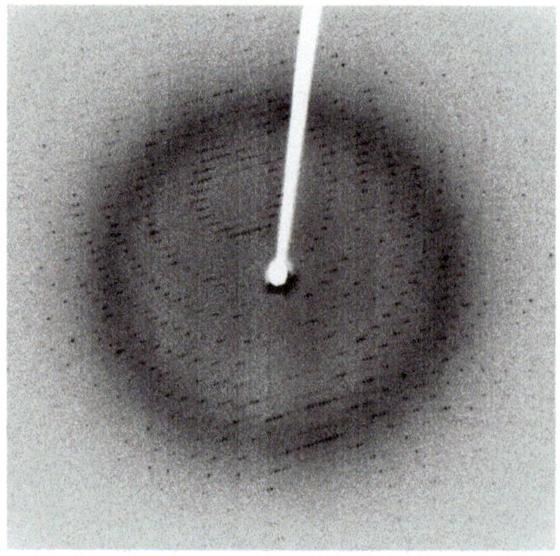

Each dot, called a reflection, in this diffraction pattern forms from the constructive interference of scattered X-rays passing through a crystal. The data can be used to determine the crystalline structure.

- X-ray crystallography in which the pattern produced by the diffraction of X-rays through the closely spaced lattice of atoms in a crystal is recorded and then analysed to reveal the nature of that lattice. A related technique, fiber diffraction, was used by Rosalind Franklin to discover the double helical structure of DNA.
- X-ray astronomy, which is an observational branch of astronomy, which deals with the study of X-ray emission from celestial objects.
- X-ray microscopic analysis, which uses electromagnetic radiation in the soft X-ray band to produce images of very small objects.
- X-ray fluorescence, a technique in which X-rays are generated within a specimen and detected. The outgoing energy of the X-ray can be used to identify the composition of the sample.
- Industrial radiography uses X-rays for inspection of industrial parts, particularly welds.

Using X-ray for inspection and quality control: the differences in the structures of the die and bond wires reveal the left chip to be counterfeit.

- Authentication and quality control, X-ray is used for authentication and quality control of packaged items.

- Industrial CT (computed tomography) is a process which uses X-ray equipment to produce three-dimensional representations of components both externally and internally. This is accomplished through computer processing of projection images of the scanned object in many directions.
- Paintings are often X-rayed to reveal underdrawings and pentimenti, alterations in the course of painting or by later restorers. Many pigments such as lead white show well in radiographs.
- X-ray spectromicroscopy has been used to analyse the reactions of pigments in paintings. For example, in analysing colour degradation in the paintings of van Gogh
- Airport security luggage scanners use X-rays for inspecting the interior of luggage for security threats before loading on aircraft.
- Border control truck scanners use X-rays for inspecting the interior of trucks.

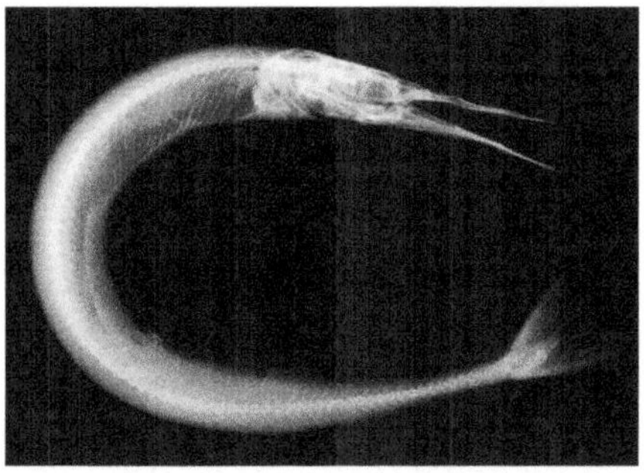

X-ray fine art photography of needlefish by Peter Dazeley

- X-ray art and fine art photography, artistic use of X-rays, for example the works by Stane Jagodič
- X-ray hair removal, a method popular in the 1920s but now banned by the FDA.
- Shoe-fitting fluoroscopes were popularized in the 1920s, banned in the US in the 1960s, banned in the UK in the 1970s, and even later in continental Europe.
- Roentgen stereophotogrammetry is used to track movement of bones based on the implantation of markers
- X-ray photoelectron spectroscopy is a chemical analysis technique relying on the photoelectric effect, usually employed in surface science.
- Radiation implosion is the use of high energy X-rays generated from a fission explosion (an A-bomb) to compress nuclear fuel to the point of fusion ignition (an H-bomb).

History

Discovery

German physicist Wilhelm Röntgen is usually credited as the discoverer of X-rays in 1895, because he was the first to systematically study them, though he is not the first to have observed their effects. He is also the one who gave them the name "X-rays" (signifying an unknown quantity) though many others referred to these as "Röntgen rays" (and the associated X-ray radiograms as, "Röntgenograms") for several decades after their discovery and even to this day in some languages, including Röntgen's native German.

Wilhelm Röntgen

Hand mit Ringen (Hand with Rings): print of Wilhelm Röntgen's first "medical" X-ray, of his wife's hand, taken on 22 December 1895 and presented to Ludwig Zehnder of the Physik Institut, University of Freiburg, on 1 January 1896

X-rays were found emanating from Crookes tubes, experimental discharge tubes invented around 1875, by scientists investigating the cathode rays, that is energetic electron beams, that were first created in the tubes. Crookes tubes created free electrons by ionization of the residual air in the tube by a high DC voltage of anywhere between a few kilovolts and 100 kV. This voltage accelerated the electrons coming from the cathode to a high enough velocity that they created X-rays when they struck the anode or the glass wall of the tube. Many of the early Crookes tubes undoubtedly radiated X-rays, because early researchers noticed effects that were attributable to them, as detailed below. Wilhelm Röntgen was the first to systematically study them, in 1895.

The discovery of X-rays stimulated a veritable sensation. Röntgen's biographer Otto Glasser estimated that, in 1896 alone, as many as 49 essays and 1044 articles about the new rays were published. This was probably a conservative estimate, if one considers that nearly every paper around the world extensively reported about the new discovery, with a magazine such as *Science* dedicating as many as 23 articles to it in that year alone. Sensationalist reactions to the new discovery included publications linking the new kind of rays to occult and paranormal theories, such as telepathy.

Early Research

Both William Crookes (in the 1880s) and German physicist Johann Hittorf, a co-inventor and early researcher of the Crookes tube, found that paper wrapped photographic plates placed near the tube became unaccountably fogged or flawed by shadows, although they had not been exposed to light. Neither found the cause nor investigated this effect.

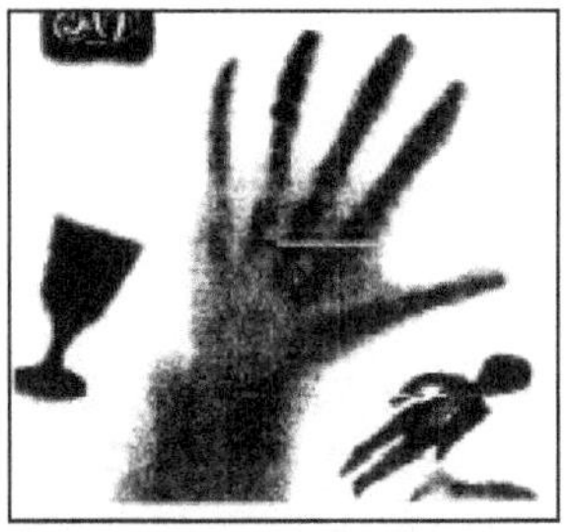

Один з перших (1895 р.) пулюєвих знімків. Рука доньки дослідника Наталі.

Один з перших пулюєвих (рентгенівських) знімків, ймовірно, якоїсь коштовної оздоби, який проф. І. Пулюй зробив до 1895 р.

Pulyui's X-ray images

In 1877 Ukrainian-born Ivan Pulyui, a lecturer in experimental physics at the University of Vienna, constructed various designs of vacuum discharge tube to investigate their properties. He continued his investigations when appointed professor at the Prague Polytechnic and in 1886 he found that sealed photographic plates became dark when exposed to the emanations from the tubes. Early in 1896, just a few weeks after Röntgen published his first X-ray photograph, Pulyui published high-quality X-ray images in journals in Paris and London. Although Pulyui had studied with Röntgen at the University of Strasbourg in the years 1873–75, his biographer Gaida (1997) asserts that his subsequent research was conducted independently.

Taking an X-ray image with early Crookes tube apparatus, late 1800s. The Crookes tube is visible in center. The standing man is viewing his hand with a fluoroscope screen. The seated man is taking a radiograph of his hand by placing it on a photographic plate. No precautions against radiation exposure are taken; its hazards were not known at the time.

X-rays were generated and detected by Fernando Sanford (1854–1948), the foundation Professor of Physics at Stanford University, in 1891. From 1886 to 1888 he had studied in the Hermann Helmholtz laboratory in Berlin, where he became familiar with the cathode rays generated in vacuum tubes when a voltage was applied across separate electrodes, as previously studied by Heinrich Hertz and Philipp Lenard. His letter of January 6, 1893 (describing his discovery as "electric photography") to The Physical Review was duly published and an article entitled *Without Lens or Light, Photographs Taken With Plate and Object in Darkness* appeared in the San Francisco Examiner.

Starting in 1888, Philipp Lenard, a student of Heinrich Hertz, conducted experiments to see whether cathode rays could pass out of the Crookes tube into the air. He built a Crookes tube (later called a "Lenard tube") with a "window" in the end made of thin aluminum, facing the cathode so the cathode rays would strike it. He found that something came through, that would expose photographic plates and cause fluorescence. He measured the penetrating power of these rays through various materials. It has been suggested that at least some of these "Lenard rays" were actually X-rays.

Hermann von Helmholtz formulated mathematical equations for X-rays. He postulated a dispersion theory before Röntgen made his discovery and announcement. It was formed on the basis of the electromagnetic theory of light. However, he did not work with actual X-rays.

In 1894 Nikola Tesla noticed damaged film in his lab that seemed to be associated with Crookes tube experiments and began investigating this *radiant energy of "invisible" kinds*. After Röntgen identified the x-ray Tesla began making X-ray images of his own using high voltages and tubes of his own design, as well as Crookes tubes.

Wilhelm Röntgen

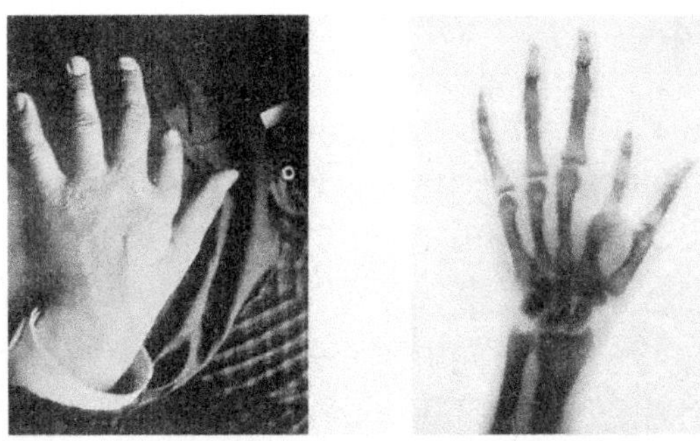

1896 plaque published in *"Nouvelle Iconographie de la Salpetrière"*, a medical journal. In the left a hand deformity, in the right same hand seen using radiography. The authors designated the technique as Röntgen photography.

On November 8, 1895, German physics professor Wilhelm Röntgen stumbled on X-rays while experimenting with Lenard and Crookes tubes and began studying them. He wrote an initial report "*On a new kind of ray: A preliminary communication*" and on December 28, 1895 submitted it to the Würzburg's Physical-Medical Society journal. This was the first paper written on X-rays. Röntgen referred to the radiation as "X", to indicate that it was an unknown type of radiation. The name stuck, although (over Röntgen's great objections) many of his colleagues suggested calling them Röntgen rays. They are still referred to as such in many languages, including German, Danish, Polish, Swedish, Finnish, Estonian, Russian, Japanese, Dutch, and Norwegian. Röntgen received the first Nobel Prize in Physics for his discovery.

There are conflicting accounts of his discovery because Röntgen had his lab notes burned after his death, but this is a likely reconstruction by his biographers: Röntgen was investigating cathode rays from a Crookes tube which he had wrapped in black cardboard so that the visible light from the tube would not interfere, using a fluorescent screen painted with barium platinocyanide. He noticed a faint green glow from the screen, about 1 meter away. Röntgen realized some invisible rays coming from the tube were passing through the cardboard to make the screen glow. He found they could also pass through books and papers on his desk. Röntgen threw himself into investigating these unknown rays systematically. Two months after his initial discovery, he published his paper.

Röntgen discovered their medical use when he made a picture of his wife's hand on a photographic plate formed due to X-rays. The photograph of his wife's hand was the first photograph of a human body part using X-rays.

Advances in Radiology

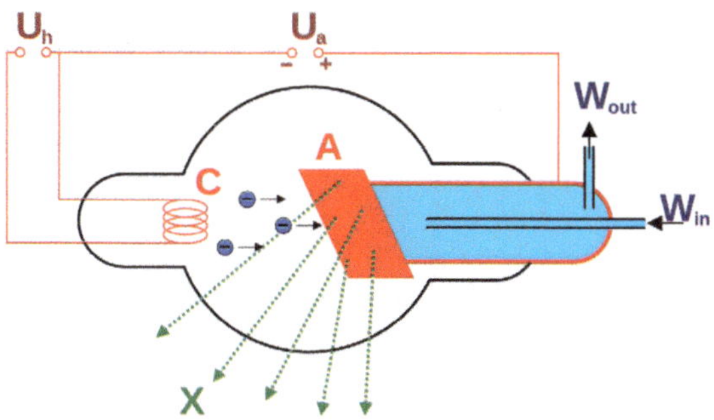

A simplified diagram of a water-cooled X-ray tube

In 1895, Thomas Edison investigated materials' ability to fluoresce when exposed to X-rays, and found that calcium tungstate was the most effective substance. Around March 1896, the fluoroscope he developed became the standard for medical X-ray examinations. Nevertheless, Edison dropped X-ray research around 1903, even before the death of Clarence Madison Dally, one of his glassblowers. Dally had a habit of testing X-ray tubes on his hands, and acquired a cancer in them so tenacious that both arms were amputated in a futile attempt to save his life.

The first use of X-rays under clinical conditions was by John Hall-Edwards in Birmingham, England on 11 January 1896, when he radiographed a needle stuck in the hand of an associate. On 14 February 1896 Hall-Edwards was also the first to use X-rays in a surgical operation. In early 1896, several weeks after Röntgen's discovery, Ivan Romanovich Tarkhanov irradiated frogs and insects with X-rays, concluding that the rays "not only photograph, but also affect the living function".

The first medical X-ray made in the United States was obtained using a discharge tube of Pulyui's design. In January 1896, on reading of Röntgen's discovery, Frank Austin of Dartmouth College tested all of the discharge tubes in the physics laboratory and found that only the Pulyui tube produced X-rays. This was a result of Pulyui's inclusion of an oblique "target" of mica, used for holding samples of fluorescent material, within the tube. On 3 February 1896 Gilman Frost, professor of medicine at the college, and his brother Edwin Frost, professor of physics, exposed the wrist of Eddie McCarthy, whom Gilman had treated some weeks earlier for a fracture, to the X-rays and collected the resulting image of the broken bone on gelatin photographic plates obtained from Howard Langill, a local photographer also interested in Röntgen's work.

In 1901, U.S. President William McKinley was shot twice in an assassination attempt. While one bullet only grazed his sternum, another had lodged somewhere deep inside his abdomen and could not be found. "A worried McKinley aide sent word to inventor Thomas Edison to rush an X-ray machine to Buffalo to find the stray bullet. It arrived but wasn't used." While the shooting itself had not been lethal, "gangrene had developed along the path of the bullet, and McKinley died of septic shock due to bacterial infection" six days later.

Hazards Discovered

With the widespread experimentation with xrays after their discovery in 1895 by scientists, physicians, and inventors came many stories of burns, hair loss, and worse in technical journals of the time. In February 1896, Professor John Daniel and Dr. William Lofland Dudley of Vanderbilt University reported hair loss after Dr. Dudley was X-rayed. A child who had been shot in the head was brought to the Vanderbilt laboratory in 1896. Before trying to find the bullet an experiment was attempted, for which Dudley "with his characteristic devotion to science" volunteered. Daniel reported that 21 days after taking a picture of Dudley's skull (with an exposure time of one hour), he noticed a bald spot 2 inches (5.1 cm) in diameter on the part of his head nearest the X-ray tube: "A plate holder with the plates towards the side of the skull was fastened and a coin placed between the skull and the head. The tube was fastened at the other side at a distance of one-half inch from the hair."

In August 1896 Dr. HD. Hawks, a graduate of Columbia College, suffered severe hand and chest burns from an x-ray demonstration. It was reported in *Electrical Review* and led to many other reports of problems associated with x-rays being sent in to the publication. Many experimenters including Elihu Thomson at Edison's lab, William J. Morton, and Nikola Tesla also reported burns. Elihu Thomson deliberately exposed a finger to an x-ray tube over a period of time and suffered pain, swelling, and blistering. Other effects were sometime blamed for the damage including ultraviolet rays and (according to Tesla) ozone. Many physicians claimed there were no effects from x-ray exposure at all. On 3 August 1905 at San Francisco, California, Elizabeth Fleischman, American woman X-ray pioneer, died from complications as a result of her work with X-rays.

20th Century and Beyond

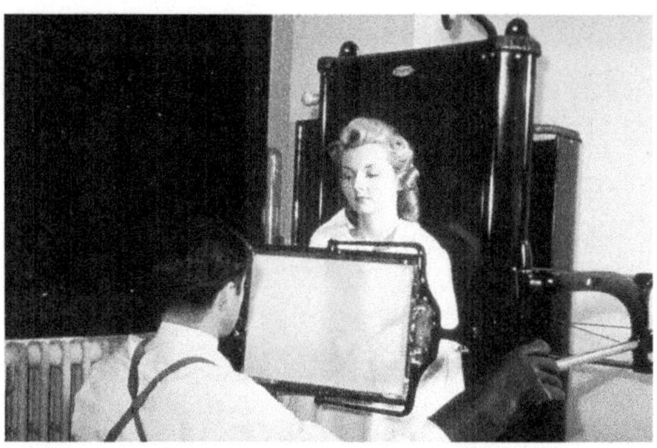

A patient being examined with a thoracic fluoroscope in 1940, which displayed continuous moving images. This image was used to argue that radiation exposure during the X-ray procedure would be negligible.

The many applications of X-rays immediately generated enormous interest. Workshops began making specialized versions of Crookes tubes for generating X-rays and these first-generation cold cathode or Crookes X-ray tubes were used until about 1920.

Crookes tubes were unreliable. They had to contain a small quantity of gas (invariably air) as a current will not flow in such a tube if they are fully evacuated. However, as time passed, the X-rays caused the glass to absorb the gas, causing the tube to generate "harder" X-rays until it soon stopped operating. Larger and more frequently used tubes were provided with devices for restoring the air, known as "softeners". These often took the form of a small side tube which contained a small piece of mica, a mineral that traps relatively large quantities of air within its structure. A small electrical heater heated the mica, causing it to release a small amount of air, thus restoring the tube's efficiency. However, the mica had a limited life, and the restoration process was difficult to control.

In 1904, John Ambrose Fleming invented the thermionic diode, the first kind of vacuum tube. This used a hot cathode that caused an electric current to flow in a vacuum. This idea was quickly applied to X-ray tubes, and hence heated-cathode X-ray tubes, called "Coolidge tubes", completely replaced the troublesome cold cathode tubes by about 1920.

In about 1906, the physicist Charles Barkla discovered that X-rays could be scattered by gases, and that each element had a characteristic X-ray spectrum. He won the 1917 Nobel Prize in Physics for this discovery.

In 1912, Max von Laue, Paul Knipping, and Walter Friedrich first observed the diffraction of X-rays by crystals. This discovery, along with the early work of Paul Peter Ewald, William Henry Bragg, and William Lawrence Bragg, gave birth to the field of X-ray crystallography.

The Coolidge X-ray tube was invented during the following year by William D. Coolidge. It made possible the continuous emissions of X-rays. X-ray tubes similar to this are still in use in 2012.

Chandra's image of the galaxy cluster Abell 2125 reveals a complex of several massive multimillion-degree-Celsius gas clouds in the process of merging.

The use of X-rays for medical purposes (which developed into the field of radiation therapy) was pioneered by Major John Hall-Edwards in Birmingham, England. Then in 1908, he had to have his left arm amputated because of the spread of X-ray dermatitis on his arm.

The X-ray microscope was developed during the 1950s.

The Chandra X-ray Observatory, launched on July 23, 1999, has been allowing the exploration of the very violent processes in the universe which produce X-rays. Unlike visible light, which gives a relatively stable view of the universe, the X-ray universe is unstable. It features stars being torn apart by black holes, galactic collisions, and novae, and neutron stars that build up layers of plasma that then explode into space.

An X-ray laser device was proposed as part of the Reagan Administration's Strategic Defense Initiative in the 1980s, but the only test of the device (a sort of laser "blaster" or death ray, powered by a thermonuclear explosion) gave inconclusive results. For technical and political reasons, the overall project (including the X-ray laser) was de-funded (though was later revived by the second Bush Administration as National Missile Defense using different technologies).

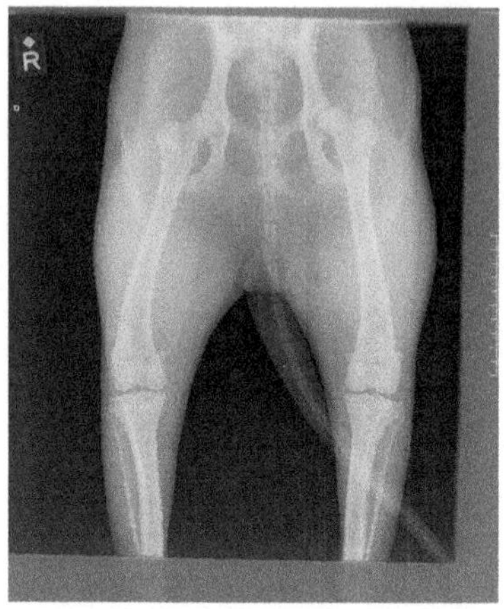

Dog hip xray posterior view

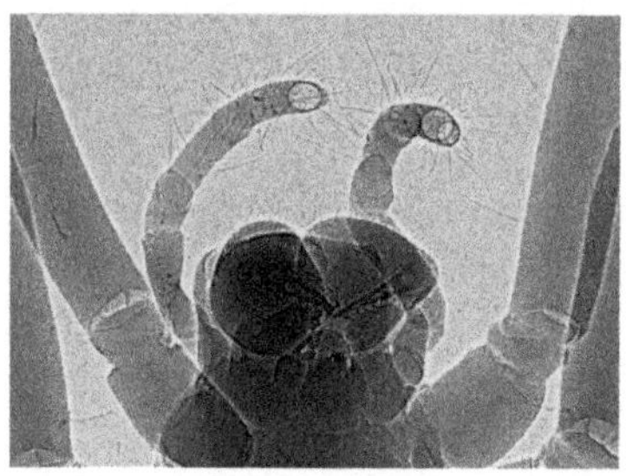

Phase-contrast x-ray image of spider

Phase-contrast X-ray imaging refers to a variety of techniques that use phase information of a coherent x-ray beam to image soft tissues. It has become an important method for visualizing cellular and histological structures in a wide range of biological and medical studies. There are several technologies being used for x-ray phase-contrast imaging, all utilizing different principles to convert phase variations in the x-rays emerging from an object into intensity variations. These include propagation-based phase contrast, talbot interferometry, refraction-enhanced imaging, and x-ray interferometry. These methods provide higher contrast compared to normal absorption-contrast x-ray imaging, making it possible to see smaller details. A disadvantage is that these methods require more sophisticated equipment, such as synchrotron or microfocus x-ray sources, X-ray optics, and high resolution x-ray detectors.

Visibility

While generally considered invisible to the human eye, in special circumstances X-rays can be visible. Brandes, in an experiment a short time after Röntgen's landmark 1895 paper, reported after dark adaptation and placing his eye close to an X-ray tube, seeing a faint "blue-gray" glow which seemed to originate within the eye itself. Upon hearing this, Röntgen reviewed his record books and found he too had seen the effect. When placing an X-ray tube on the opposite side of a wooden door Röntgen had noted the same blue glow, seeming to emanate from the eye itself, but thought his observations to be spurious because he only saw the effect when he used one type of tube. Later he realized that the tube which had created the effect was the only one powerful enough to make the glow plainly visible and the experiment was thereafter readily repeatable. The knowledge that X-rays are actually faintly visible to the dark-adapted naked eye has largely been forgotten today; this is probably due to the desire not to repeat what would now be seen as a recklessly dangerous and potentially harmful experiment with ionizing radiation. It is not known what exact mechanism in the eye produces the visibility: it could be due to conventional detection (excitation of rhodopsin molecules in the retina), direct excitation of retinal nerve cells, or secondary detection via, for instance, X-ray induction of phosphorescence in the eyeball with conventional retinal detection of the secondarily produced visible light.

Though X-rays are otherwise invisible, it is possible to see the ionization of the air molecules if the intensity of the X-ray beam is high enough. The beamline from the wiggler at the ID11 at the European Synchrotron Radiation Facility is one example of such high intensity.

Units of Measure and Exposure

The measure of X-rays ionizing ability is called the exposure:

- The coulomb per kilogram (C/kg) is the SI unit of ionizing radiation exposure, and it is the amount of radiation required to create one coulomb of charge of each polarity in one kilogram of matter.
- The roentgen (R) is an obsolete traditional unit of exposure, which represented the amount of radiation required to create one electrostatic unit of charge of each polarity in one cubic centimeter of dry air. 1 roentgen= 2.58×10^{-4} C/kg.

However, the effect of ionizing radiation on matter (especially living tissue) is more closely related to the amount of energy deposited into them rather than the charge generated. This measure of energy absorbed is called the absorbed dose:

- The gray (Gy), which has units of (joules/kilogram), is the SI unit of absorbed dose, and it is the amount of radiation required to deposit one joule of energy in one kilogram of any kind of matter.
- The rad is the (obsolete) corresponding traditional unit, equal to 10 millijoules of energy deposited per kilogram. 100 rad= 1 gray.

The equivalent dose is the measure of the biological effect of radiation on human tissue. For X-rays it is equal to the absorbed dose.

- The Roentgen equivalent man (rem) is the traditional unit of equivalent dose. For X-rays it is equal to the rad, or, in other words, 10 millijoules of energy deposited per kilogram. 100 rem = 1 Sv.
- The sievert (Sv) is the SI unit of equivalent dose, and also of effective dose. For X-rays the "equivalent dose" is numerically equal to a Gray (Gy). 1 Sv= 1 Gy. For the "effective dose" of X-rays, it is usually not equal to the Gray (Gy).

Diffraction Methods

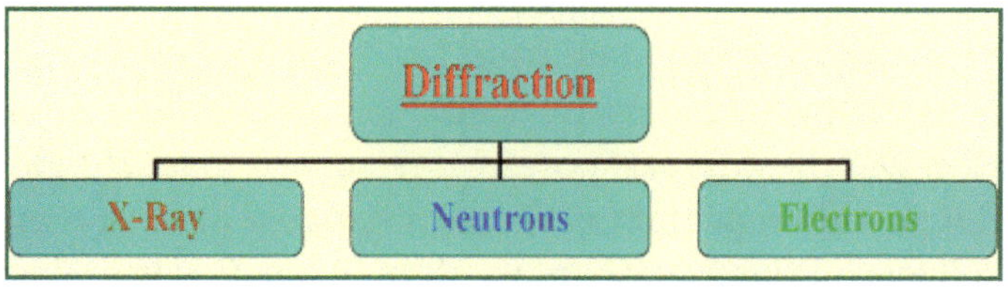

- Radiation sources other than X-Rays such as neutrons or electrons can also be used in crystal diffraction experiments.
- The physical basis for the diffraction of electrons or neutrons is the same as that for the diffraction of X-Rays. The only difference is in the mechanism of scattering. The typical wavelength and energy involved for various sources are given below:

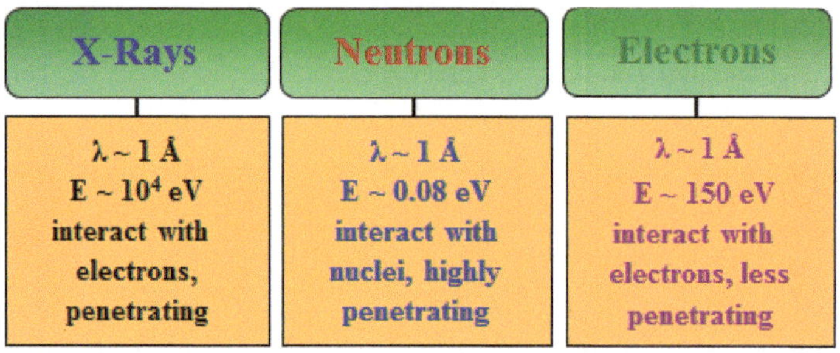

Wavelength Range of X-rays in Electromagnetic Spectrum

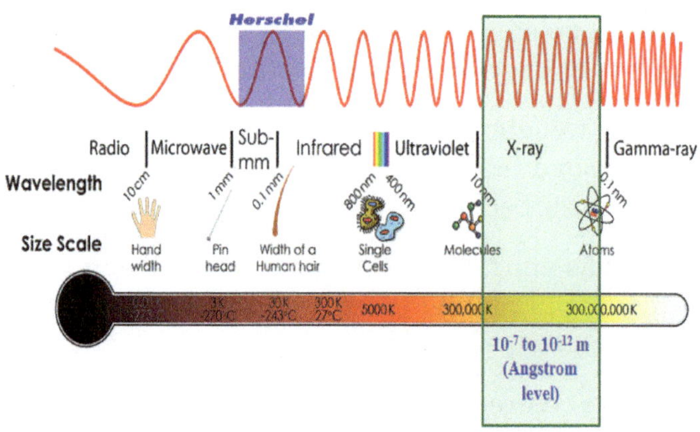

Spectral range of electromagnetic waves.

Figure depicts the spectral range of electromagnetic waves. The wavelength of X-rays is longer than gamma rays but shorter than ultraviolet rays. These rays are widely used for diagnostic radiography and crystallography, covering an extremely wide area.

Generation of X-rays

In general, X-rays are generated either by X-ray tube or by synchrotron radiation. The X-ray tube is the primary X-ray source used in laboratories, whereas the synchrotron radiation needs a large-scale accelerator for obtaining high-energy radiation.

Figure shown below demonstrates the X-rays induced by colliding high-energy thermal electrons with a metallic target (water cooled) such as copper, molybdenum, etc., which can emit strong X-rays of 1.541 Å, 0.71 Å, respectively. The emission of different X-rays from the copper metal is also shown schematically in figure.

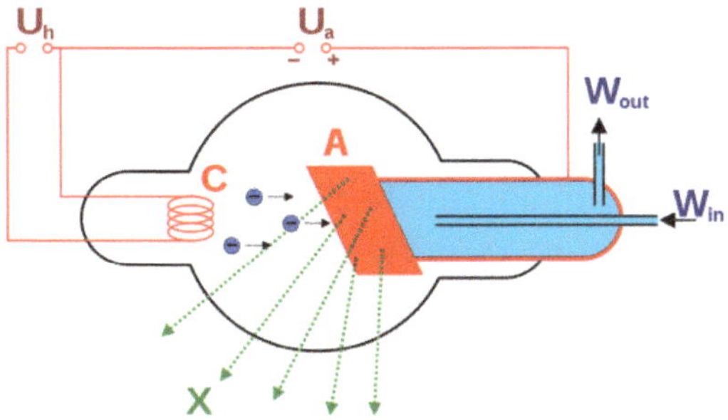

Schematic diagram of a water cooled X-ray tube.

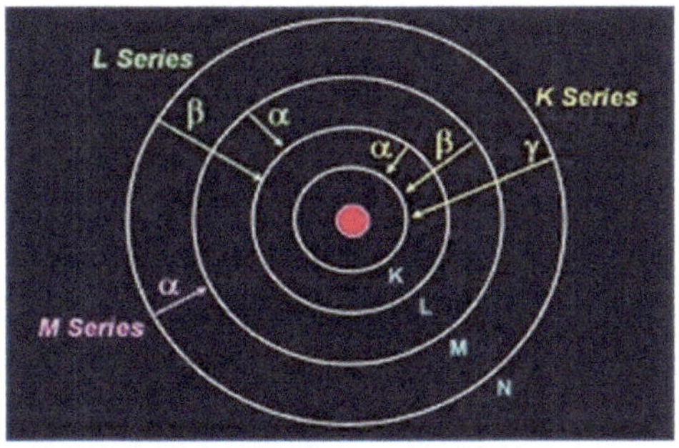

The characteristic X-rays are generated when the electron in the shells is flipped by high energy thermal electrons emitted from a filament.

Diffraction Basics

Father Sir William Henry Bragg (left) and Son Sir William Lawrence Bragg (right)

Father Sir William Henry Bragg and Son Sir William Lawrence Bragg proposed that Bragg's diffraction occurs when electromagnetic radiation or subatomic particle wave with a wavelength comparable to atomic spacing is incident upon a crystalline sample and scattered in a specular fashion by the atoms in the system and undergoes a constructive interference.

For a crystalline solid, the waves are scattered from lattice planes separated by the inter-planar distance d. When the scattered waves interfere constructively, they remain in phase (demonstrated later) since the path length of each wave is equal to an integer multiple of the wavelength.

The path difference between the two waves undergoing constructive interference is given by 2d sinθ, where θ is the scattering angle. This leads to Bragg's law, which describes the condition for constructive interference from successive crystallographic planes (hkl) of the crystalline lattice.

Bragg's Law

In physics, Bragg's law, or Wulff–Bragg's condition in post-Soviet countries, a special case of Laue diffraction, gives the angles for coherent and incoherent scattering from a crystal lattice. When X-rays are incident on an atom, they make the electronic cloud move as does any electromagnetic wave. The movement of these charges re-radiates waves with the same frequency, blurred slightly due to a variety of effects; this phenomenon is known as Rayleigh scattering (or elastic scattering). The scattered waves can themselves be scattered but this secondary scattering is assumed to be negligible.

A similar process occurs upon scattering neutron waves from the nuclei or by a coherent spin interaction with an unpaired electron. These re-emitted wave fields interfere with each other either constructively or destructively (overlapping waves either add up together to produce stronger peaks or are subtracted from each other to some degree), producing a diffraction pattern on a detector or film. The resulting wave interference pattern is the basis of diffraction analysis. This analysis is called *Bragg diffraction*.

History

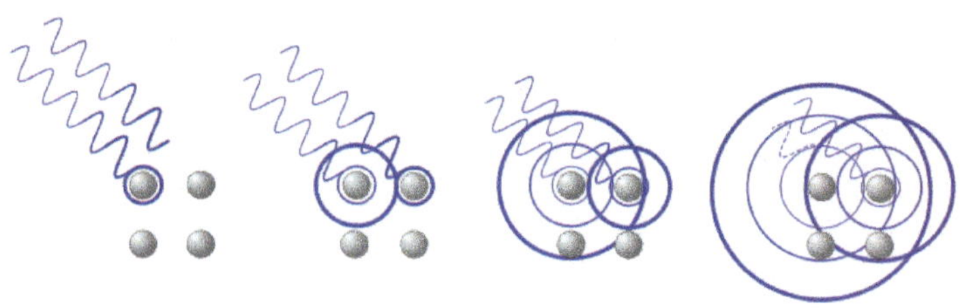

X-rays interact with the atoms in a crystal.

Bragg diffraction (also referred to as the Bragg formulation of X-ray diffraction) was first proposed by William Lawrence Bragg and his father William Henry Bragg in 1913 in response to their discovery that crystalline solids produced surprising patterns of reflected X-rays (in contrast to that of, say, a liquid). They found that these crystals, at certain specific wavelengths and incident angles, produced intense peaks of reflected radiation. The concept of Bragg diffraction applies equally to neutron diffraction and electron diffraction processes. Both neutron and X-ray wavelengths are comparable with inter-atomic distances (~ 150 pm) and thus are an excellent probe for this length scale.

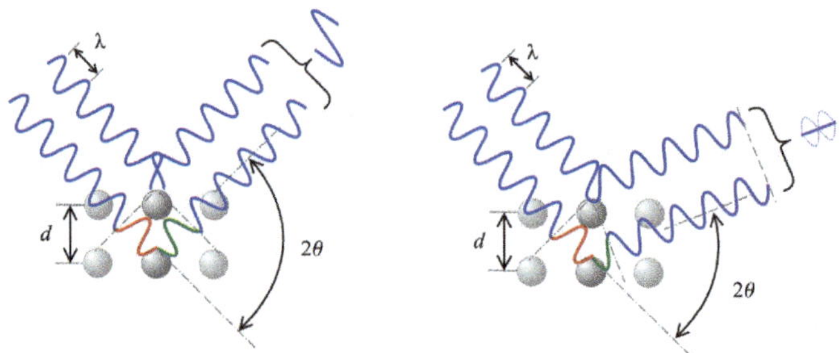

According to the 2θ deviation, the phase shift causes constructive (left figure) or destructive (right figure) interferences.

Sir William Lawrence Bragg explained this result by modeling the crystal as a set of discrete parallel planes separated by a constant parameter *d*. It was proposed that the incident X-ray radiation would produce a Bragg peak if their reflections off the various planes interfered constructively. The interference is constructive when the phase shift is a multiple of 2π; this condition can be expressed by Bragg's law and was first presented by Sir William Lawrence Bragg on 11 November 1912 to the Cambridge Philosophical Society. Although simple, Bragg's law confirmed the existence of real particles at the atomic scale, as well as providing a powerful new tool for studying crystals in the form of X-ray and neutron diffraction. William Lawrence Bragg and his father, Sir William Henry Bragg, were awarded the Nobel Prize in physics in 1915 for their work in determining crystal structures beginning with NaCl, ZnS, and diamond. They are the only father-son team to jointly win. William Lawrence Bragg was 25 years old, making him then, the youngest physics Nobel laureate.

Bragg Condition

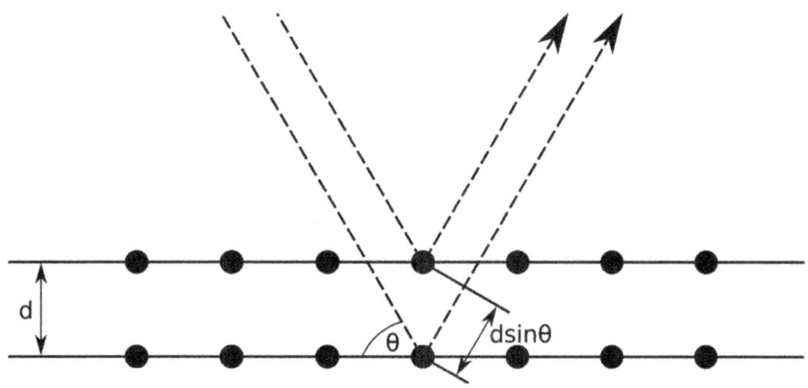

Bragg diffraction. Two beams with identical wavelength and phase approach a crystalline solid and are scattered off two different atoms within it. The lower beam traverses an extra length of $2d\sin\theta$. Constructive interference occurs when this length is equal to an integer multiple of the wavelength of the radiation.

Bragg diffraction occurs when radiation, with wavelength comparable to atomic spacings, is scattered in a specular fashion by the atoms of a crystalline system, and undergoes constructive interference. For a crystalline solid, the waves are scattered from lattice planes separated by the interplanar distance *d*. When the scattered waves interfere constructively, they remain in phase since the difference between the path lengths of the two waves is equal to an integer multiple of the wavelength. The path difference between two waves undergoing interference is given by $2d\sin\theta$, where θ is the scattering angle. The effect of the constructive or destructive interference intensifies because of the cumulative effect of reflection in successive crystallographic planes of the crystalline lattice (as described by Miller notation). This leads to Bragg's law, which describes the condition on θ for the constructive interference to be at its strongest:

$$2d\sin\theta = n\lambda,$$

where n is a positive integer and λ is the wavelength of incident wave. Note that moving particles, including electrons, protons and neutrons, have an associated wavelength called *de Broglie wavelength*. A diffraction pattern is obtained by measuring the intensity of scattered waves as a function of scattering angle. Very strong intensities known as Bragg peaks are obtained in the diffraction pattern at the points where the scattering angles satisfy Bragg condition.

Heuristic Derivation

Suppose that a single monochromatic wave (of any type) is incident on aligned planes of lattice points, with separation d, at angle θ. Points A and C are on one plane, and B is on the plane below. Points ABCC' form a quadrilateral.

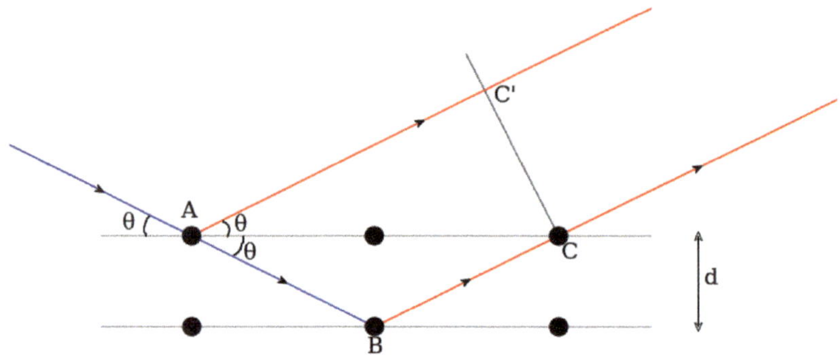

There will be a path difference between the ray that gets reflected along AC' and the ray that gets transmitted, then reflected, along AB and BC respectively. This path difference is

$$(AB + BC) - (AC').$$

The two separate waves will arrive at a point with the same phase, and hence undergo constructive interference, if and only if this path difference is equal to any integer value of the wavelength, i.e.

$$(AB + BC) - (AC') = n\lambda,$$

where the same definition of n and λ apply as above.

Therefore,

$$AB = BC = \frac{d}{\sin\theta} \text{ and } AC = \frac{2d}{\tan\theta},$$

from which it follows that

$$AC' = AC \cdot \cos\theta = \frac{2d}{\tan\theta}\cos\theta = \left(\frac{2d}{\sin\theta}\cos\theta\right)\cos\theta = \frac{2d}{\sin\theta}\cos^2\theta.$$

Putting everything together,

$$n\lambda = \frac{2d}{\sin\theta}(1 - \cos^2\theta) = \frac{2d}{\sin\theta}\sin^2\theta,$$

which simplifies to

$$n\lambda = 2d\sin\theta,$$

which is Bragg's law.

If only two planes of atoms were diffracting, as shown in the pictures, then the transition from constructive to destructive interference would be gradual as a function of angle, with gentle maxima at the Bragg angles. However, since many atomic planes are interfering in real materials, very sharp peaks surrounded by mostly destructive interference result.

Bragg Scattering of Visible Light by Colloids

A colloidal crystal is a highly ordered array of particles that forms over a long range (from a few millimeters to one centimeter in length); colloidal crystals have appearance and properties roughly analogous to their atomic or molecular counterparts. It has been known for many years that, due to repulsive Coulombic interactions, electrically charged macromolecules in an aqueous environment can exhibit long-range crystal-like correlations, with interparticle separation distances often being considerably greater than the individual particle diameter. Periodic arrays of spherical particles give rise to interstitial voids (the spaces between the particles), which act as a natural diffraction grating for visible light waves, when the interstitial spacing is of the same order of magnitude as the incident lightwave. In these cases in nature, brilliant iridescence (or play of colours) is attributed to the diffraction and constructive interference of visible lightwaves according to Bragg's law, in a matter analogous to the scattering of X-rays in crystalline solid. The effects occur at visible wavelengths because the separation parameter d is much larger than for true crystals.

Volume Bragg Gratings

Volume Bragg Gratings (VBG) or Volume holographic gratings (VHG) consist of a volume where there is a periodic change in the refractive index. Depending on the orientation of the modulation of the refractive index, VBG can be use either to transmit or reflect a small bandwidth of wavelengths. Bragg's law (adapted for volume hologram) dictates which wavelength will be diffracted:

$$2n\Lambda\sin(\theta+\phi)=\lambda_B,$$

where n is a positive integer, λ_B the diffracted wavelength, Λ the step of the grating, θ the angle between the incident beam and the normal (N) of the entrance surface and φ the angle between the normal and the grating vector (K_G). Radiation that does not match Bragg's law will pass through the VBG undiffracted. The output wavelength can be tuned over a few hundred nanometers by changing the incident angle (θ). VBG are being used to produce widely tunable laser source or perform global hyperspectral imagery.

Selection Rules and Practical Crystallography

Bragg's law, as stated above, can be used to obtain the lattice spacing of a particular cubic system through the following relation:

$$d=\frac{a}{\sqrt{h^2+k^2+l^2}},$$

where a is the lattice spacing of the cubic crystal, and h, k, and l are the Miller indices of the Bragg plane. Combining this relation with Bragg's law:

$$\left(\frac{\lambda}{2a}\right)^2=\left(\frac{\lambda}{2d}\right)^2\frac{1}{h^2+k^2+l^2}$$

Bravais lattice	Example compounds	Allowed reflections	Forbidden reflections
Simple cubic	Po	Any h, k, l	None
Body-centered cubic	Fe, W, Ta, Cr	$h+k+l=$ even	$h+k+l=$ odd

Face-centered cubic	Cu, Al, Ni, NaCl, LiH, PbS	h, k, l all odd or all even	h, k, l mixed odd and even
Diamond F.C.C.	Si, Ge	all odd, or all even with $h+k+l = 4n$	h, k, l mixed odd and even, or all even with $h+k+l \neq 4n$
Triangular lattice	Ti, Zr, Cd, Be	l even, $h + 2k \neq 3n$	$h + 2k = 3n$ for odd l

These selection rules can be used for any crystal with the given crystal structure. KCl exhibits a fcc cubic structure. However, the K^+ and the Cl^- ion have the same number of electrons and are quite close in size, so that the diffraction pattern becomes essentially the same as for a simple cubic structure with half the lattice parameter. Selection rules for other structures can be referenced elsewhere, or derived.

$$n\lambda = 2\, d\, sin\theta$$

- Constructive interference only occurs for certain θ' s correlating to a (hkl) plane, specifically when the path difference is equal to n wavelengths, as schematically shown in figure.

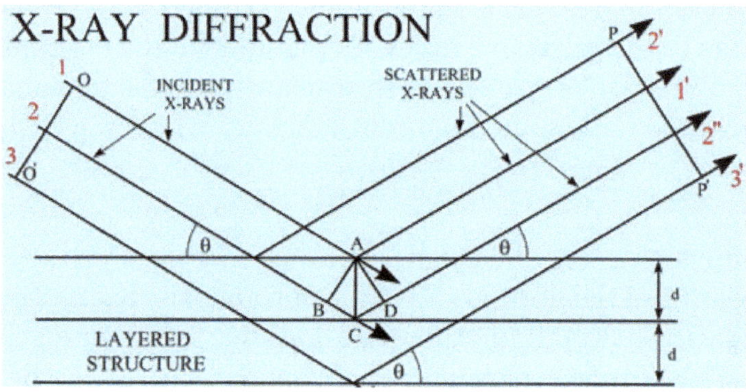

Schematic of X-ray diffraction.

The typical set up of X-ray diffractometer geometry and its components are shown in figure.

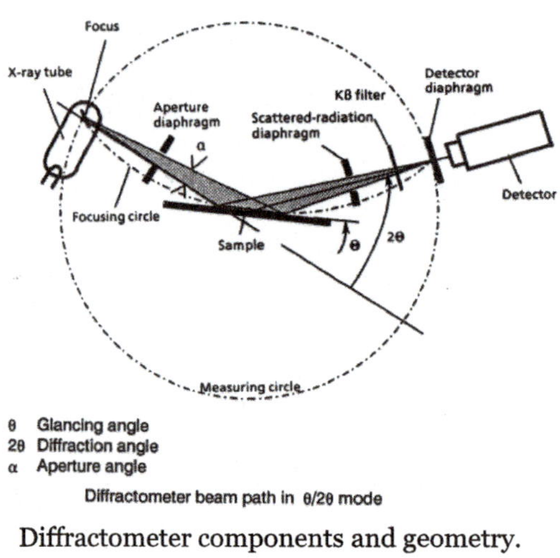

Diffractometer components and geometry.

The typical X-ray diffractometer and projected geometry are displayed in figure to study the crystal structure of the materials.

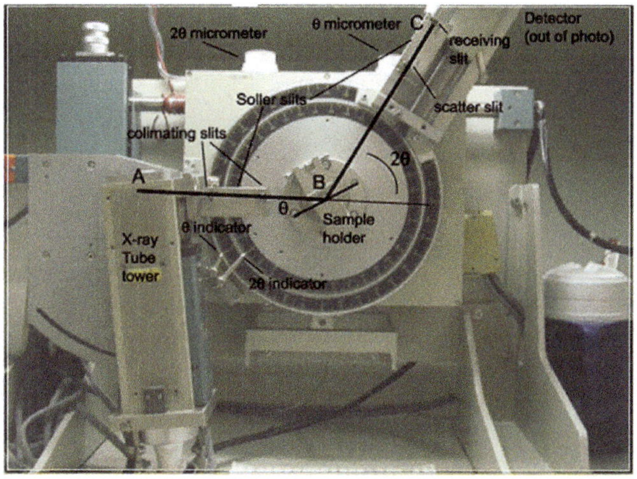

X-ray diffractometer and geometry.

Constructive Interference

Let us understand how the waves scattered from two successive planes interfering constructively. Figure demonstrates the interference of two waves constructively, when the path difference matches with wavelength.

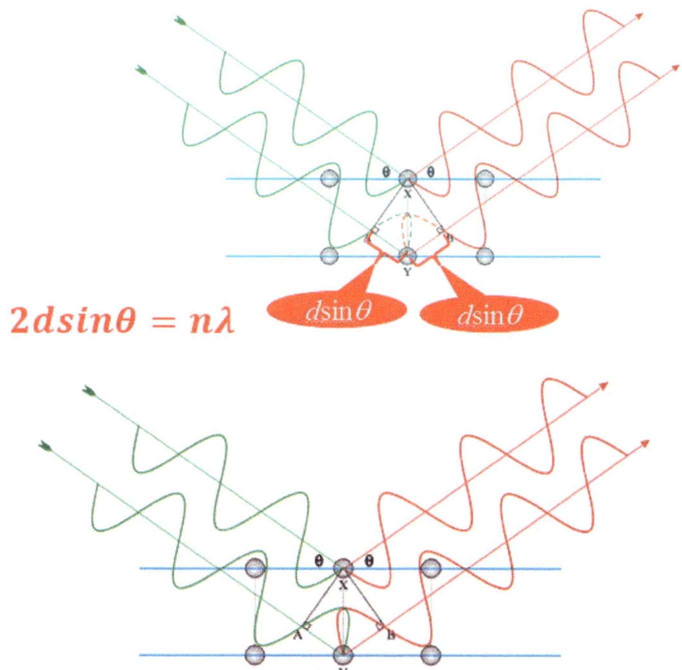

Schematic of Constructive interference

Since n = 1, the path difference should match with one wavelength.

Assuming that path difference of λ gives constructive interference: Similar to the path difference of λ, path difference of 2λ, 3λ ... nλ also constructively interfere, as displayed in figure.

All Constructive interference

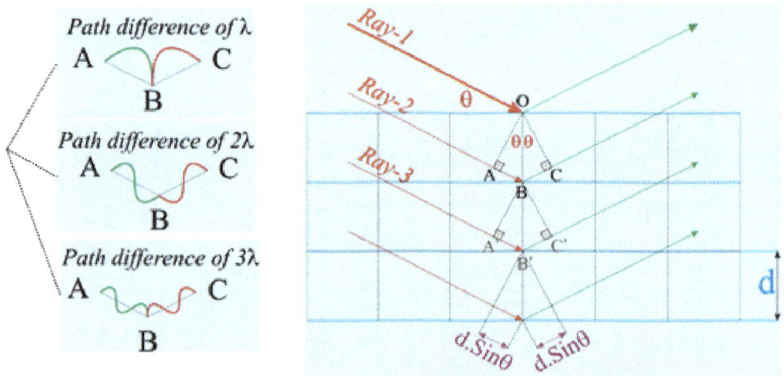

Schematic of path difference between more planes.

It may be noted that if the path difference between Ray-1 and Ray-2 is λ then the path difference between Ray-1 and Ray-3 is 2λ and Ray-1 and Ray-4 is 3λ, etc.

Similarly, one could go across different planes as shown in figure.

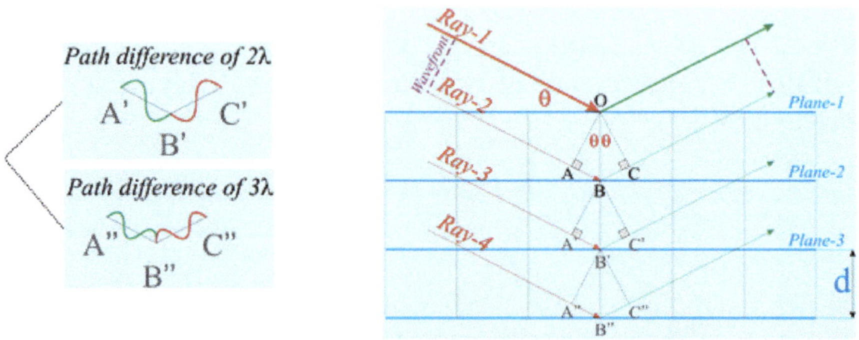

Schematic of path difference going across planes.

Interference (Wave Propagation)

The iridescence of soap bubbles is due to thin-film interference.

In physics, interference is a phenomenon in which two waves superpose to form a resultant wave of greater, lower, or the same amplitude. Interference usually refers to the interaction of waves that are correlated or coherent with each other, either because they come from the same source or because they have the same or nearly the same frequency. Interference effects can be observed with all types of waves, for example, light, radio, acoustic, surface water waves or matter waves.

Mechanisms

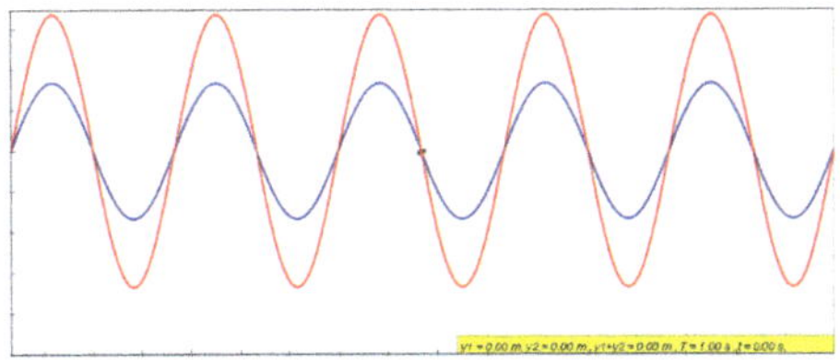

Interference of left traveling (green) and right traveling (blue) waves in one dimension, resulting in final (red) wave

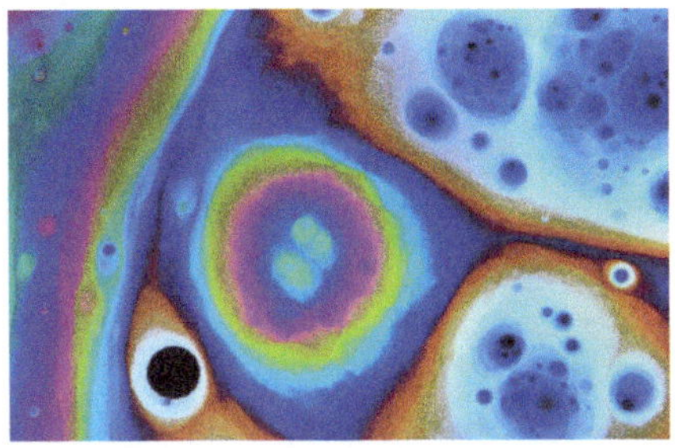

A magnified image of a coloured interference pattern in a soap film. The "black holes" are areas of almost total destructive interference, (antiphase).

Interference of waves from two point sources.

The principle of superposition of waves states that when two or more propagating waves of same type are incident on the same point, the resultant amplitude at that point is equal to the vector sum of the amplitudes of the individual waves. If a crest of a wave meets a crest of another wave of the same frequency at the same point, then the amplitude is the sum of the individual amplitudes—this is constructive interference. If a crest of one wave meets a trough of another wave, then the amplitude is equal to the difference in the individual amplitudes—this is known as destructive interference.

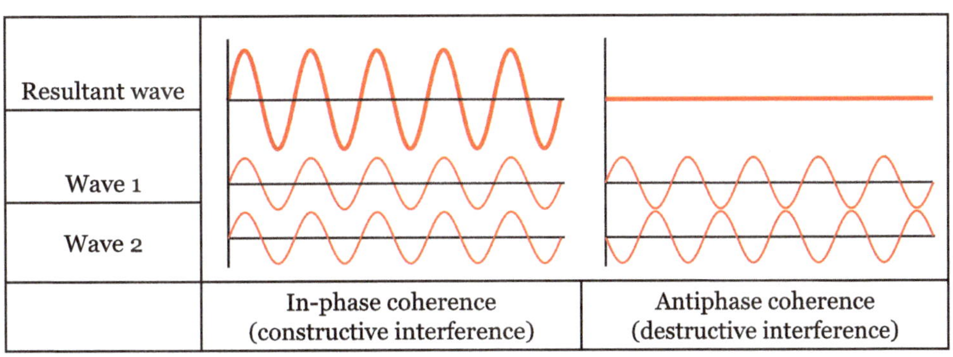

Constructive interference occurs when the phase difference between the waves is an even multiple of π (180°) (a multiple of 2π, 360°), whereas destructive interference occurs when the difference is an odd multiple of π. If the difference between the phases is intermediate between these two extremes, then the magnitude of the displacement of the summed waves lies between the minimum and maximum values.

Consider, for example, what happens when two identical stones are dropped into a still pool of water at different locations. Each stone generates a circular wave propagating outwards from the point where the stone was dropped. When the two waves overlap, the net displacement at a particular point is the sum of the displacements of the individual waves. At some points, these will be in phase, and will produce a maximum displacement. In other places, the waves will be in anti-phase, and there will be no net displacement at these points. Thus, parts of the surface will be stationary—these are seen in the figure as stationary blue-green lines radiating from the centre.

Interference of light is a common phenomenon that can be explained classically by the superposition of waves theory however a deeper understanding of light interference requires knowledge of the quantum wave propagation of light which resolves further experimental observations. The Strange Theory of Light and Matter. Prime examples of light interference are the famous Double-slit experiment, laser speckle, optical thin layers and films and interferometers. An example is the double slit experiment, classically light interferes and the energy of photons is lost, however with wave propagation the observed bright and dark areas are a result of the paths available for the photons to travel. Dark areas in the double slit are not available to the photons and bright areas are allowed paths. Laser speckle and a Michelson interferometer are examples where an observer truly observes light of differing phases, the electrons in the photo sensitive areas of the eye do not perceive photons that are out of phase with each other due to a net zero superposition of the E vector of the photons, these photons merely continue deeper into the eye tissues where they are absorbed. Thin films also behave in a quantum manner. Traditionally the classical model is taught as a basis for understanding optical interference based the Huygens–Fresnel principle and it was not until discussions in the 1920s Solvay Conference that de Broglie first proposed unique wave properties of light, Feynman made further significant contributions in the 1940s/50s and experiments continue today.

Derivation

The above can be demonstrated in one dimension by deriving the formula for the sum of two waves. The equation for the amplitude of a sinusoidal wave traveling to the right along the x-axis is

$$W_1(x,t) = A\cos(kx - \omega t)$$

where A is the peak amplitude, $k = 2\pi / \lambda$ is the wavenumber and $\omega = 2\pi f$ is the angular frequency of the wave. Suppose a second wave of the same frequency and amplitude but with a different phase is also traveling to the right

$$W_2(x,t) = A\cos(kx - \omega t + \phi)$$

where ϕ is the phase difference between the waves in radians. The two waves will superpose and add: the sum of the two waves is

$$W_1 + W_2 = A[\cos(kx - \omega t) + \cos(kx - \omega t + \phi)]$$

Using the trigonometric identity for the sum of two cosines: $\cos a + \cos b = 2\cos(\frac{a+b}{2})\cos(\frac{a-b}{2})$, this can be written

$$W_1 + W_2 = 2A\cos(kx - \omega t + \frac{\phi}{2})\cos(\frac{\phi}{2})$$

This represents a wave at the original frequency, traveling to the right like the components, whose amplitude is proportional to the cosine of $\phi / 2$.

- *Constructive interference*: If the phase difference is an even multiple of pi: $\phi = ... - 4\pi, -2\pi, 0, 2\pi, 4\pi, ...$ then $|\cos\phi / 2| = 1$, so the sum of the two waves is a wave with twice the amplitude

$$W_1 + W_2 = 2A\cos(kx - \omega t)$$

- *Destructive interference*: If the phase difference is an odd multiple of pi: $\phi = ... - 3\pi, -\pi, \pi, 3\pi, 5\pi, ...$ then $\cos\phi / 2 = 1$, so the sum of the two waves is zero

$$W_1 + W_2 = 0$$

Between two Plane Waves

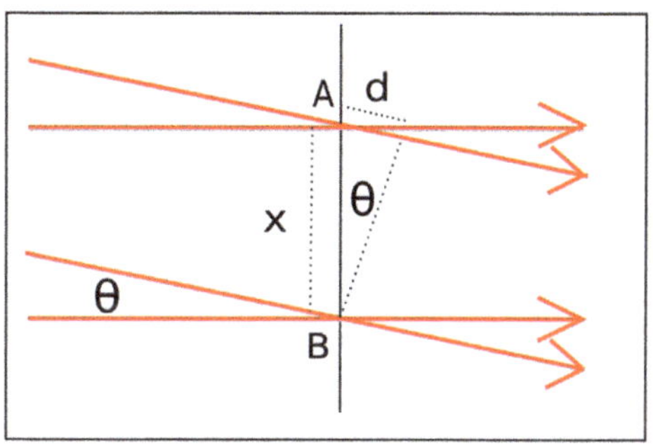

Geometrical arrangement for two plane wave interference

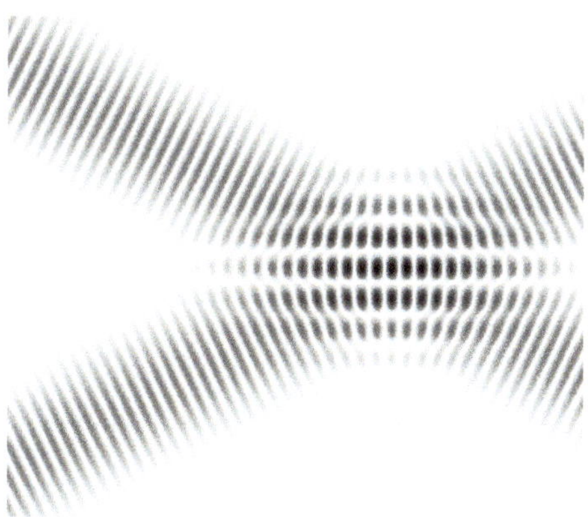

Interference fringes in overlapping plane waves

A simple form of interference pattern is obtained if two plane waves of the same frequency intersect at an angle. Interference is essentially an energy redistribution process. The energy which is lost at the destructive interference is regained at the constructive interference. One wave is travelling horizontally, and the other is travelling downwards at an angle θ to the first wave. Assuming that the two waves are in phase at the point B, then the relative phase changes along the x-axis. The phase difference at the point A is given by

$$\Delta\varphi = \frac{2\pi d}{\lambda} = \frac{2\pi x \sin\theta}{\lambda}$$

It can be seen that the two waves are in phase when

$$\frac{x\sin\theta}{\lambda} = 0, \pm 1, \pm 2, \ldots,$$

and are half a cycle out of phase when

$$\frac{x\sin\theta}{\lambda} = \pm\frac{1}{2}, \pm\frac{3}{2}, \ldots$$

Constructive interference occurs when the waves are in phase, and destructive interference when they are half a cycle out of phase. Thus, an interference fringe pattern is produced, where the separation of the maxima is

$$d_f = \frac{\lambda}{\sin\theta}$$

and d_f is known as the fringe spacing. The fringe spacing increases with increase in wavelength, and with decreasing angle θ.

The fringes are observed wherever the two waves overlap and the fringe spacing is uniform throughout.

Between Two Spherical Waves

Optical interference between two point sources for different wavelengths and source separations

A point source produces a spherical wave. If the light from two point sources overlaps, the interference pattern maps out the way in which the phase difference between the two waves varies in space. This depends on the wavelength and on the separation of the point sources. The figure to the right shows interference between two spherical waves. The wavelength increases from top to bottom, and the distance between the sources increases from left to right.

When the plane of observation is far enough away, the fringe pattern will be a series of almost straight lines, since the waves will then be almost planar.

Multiple Beams

Interference occurs when several waves are added together provided that the phase differences between them remain constant over the observation time.

It is sometimes desirable for several waves of the same frequency and amplitude to sum to zero (that is, interfere destructively, cancel). This is the principle behind, for example, 3-phase power and the diffraction grating. In both of these cases, the result is achieved by uniform spacing of the phases.

It is easy to see that a set of waves will cancel if they have the same amplitude and their phases are spaced equally in angle. Using phasors, each wave can be represented as $Ae^{i\varphi_n}$ for N waves from $n=0$ to $n=N-1$, , where

$$\varphi_n - \varphi_{n-1} = \frac{2\pi}{N}.$$

To show that

$$\sum_{n=0}^{N-1} Ae^{i\varphi_n} = 0$$

one merely assumes the converse, then multiplies both sides by $e^{i\frac{2\pi}{N}}$

The Fabry–Pérot interferometer uses interference between multiple reflections.

A diffraction grating can be considered to be a multiple-beam interferometer, since the peaks which it produces are generated by interference between the light transmitted by each of the elements in the grating.

Optical Interference

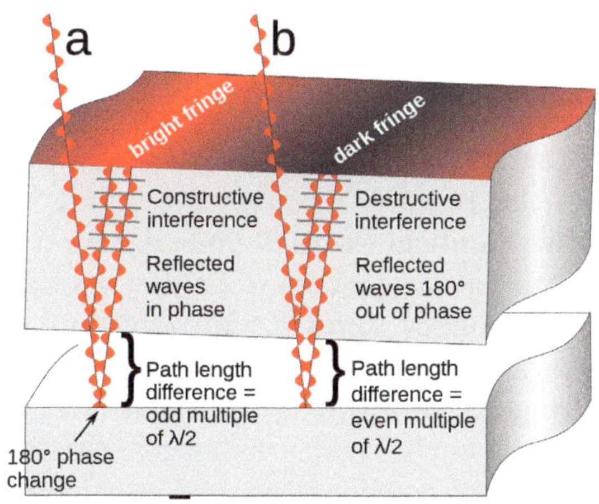

Creation of interference fringes by an optical flat on a reflective surface. Light rays from a monochromatic source pass through the glass and reflect off both the bottom surface of the flat and the supporting surface. The tiny gap between the surfaces means the two reflected rays have different path lengths. In addition the ray reflected from the bottom plate undergoes a 180° phase reversal. As a result, at locations *(a)* where the path difference is an odd multiple of λ/2, the waves reinforce. At locations *(b)* where the path difference is an even multiple of λ/2 the waves cancel. Since the gap between the surfaces varies slightly in width at different points, a series of alternating bright and dark bands, *interference fringes*, are seen.

Because the frequency of light waves (~10^{14} Hz) is too high to be detected by currently available detectors, it is possible to observe only the intensity of an optical interference pattern. The intensity of the light at a given point is proportional to the square of the average amplitude of the wave. This can be expressed mathematically as follows. The displacement of the two waves at a point r is:

$$U_1(\mathbf{r},t) = A_1(\mathbf{r})e^{i[\varphi_1(\mathbf{r})-\omega t]}$$

$$U_2(\mathbf{r},t) = A_2(\mathbf{r})e^{i[\varphi_2(\mathbf{r})-\omega t]}$$

where A represents the magnitude of the displacement, φ represents the phase and ω represents the angular frequency.

The displacement of the summed waves is

$$U(\mathbf{r},t) = A_1(\mathbf{r})e^{i[\varphi_1(\mathbf{r})-\omega t]} + A_2(\mathbf{r})e^{i[\varphi_2(\mathbf{r})-\omega t]}$$

The intensity of the light at r is given by

$$I(\mathbf{r}) = \int U(\mathbf{r},t)U^*(\mathbf{r},t)dt \propto A_1^2(\mathbf{r}) + A_2^2(\mathbf{r}) + 2A_1(\mathbf{r})A_2(\mathbf{r})\cos[\varphi_1(\mathbf{r}) - \varphi_2(\mathbf{r})]$$

This can be expressed in terms of the intensities of the individual waves as

$$I(\mathbf{r}) = I_1(\mathbf{r}) + I_2(\mathbf{r}) + 2\sqrt{I_1(\mathbf{r})I_2(\mathbf{r})}\cos[\varphi_1(\mathbf{r}) - \varphi_2(\mathbf{r})]$$

Thus, the interference pattern maps out the difference in phase between the two waves, with maxima occurring when the phase difference is a multiple of 2π. If the two beams are of equal intensity, the maxima are four times as bright as the individual beams, and the minima have zero intensity.

The two waves must have the same polarization to give rise to interference fringes since it is not possible for waves of different polarizations to cancel one another out or add together. Instead, when waves of different polarization are added together, they give rise to a wave of a different polarization state.

Light Source Requirements

The discussion above assumes that the waves which interfere with one another are monochromatic, i.e. have a single frequency—this requires that they are infinite in time. This is not, however, either practical or necessary. Two identical waves of finite duration whose frequency is fixed over that period will give rise to an interference pattern while they overlap. Two identical waves which consist of a narrow spectrum of frequency waves of finite duration, will give a series of fringe patterns of slightly differing spacings, and provided the spread of spacings is significantly less than the average fringe spacing, a fringe pattern will again be observed during the time when the two waves overlap.

Conventional light sources emit waves of differing frequencies and at different times from different points in the source. If the light is split into two waves and then re-combined, each individual light wave may generate an interference pattern with its other half, but the individual fringe patterns generated will have different phases and spacings, and normally no overall fringe pattern will be observable. However, single-element light sources, such as sodium- or mercury-vapor lamps have emission lines with quite narrow frequency spectra. When these are spatially and colour filtered, and then split into two waves, they can be superimposed to generate interference fringes. All interferometry prior to the invention of the laser was done using such sources and had a wide range of successful applications.

A laser beam generally approximates much more closely to a monochromatic source, and it is much more straightforward to generate interference fringes using a laser. The ease with which interference fringes can be observed with a laser beam can sometimes cause problems in that stray reflections may give spurious interference fringes which can result in errors.

Normally, a single laser beam is used in interferometry, though interference has been observed using two independent lasers whose frequencies were sufficiently matched to satisfy the phase requirements.

White light interference in a soap bubble

It is also possible to observe interference fringes using white light. A white light fringe pattern can be considered to be made up of a 'spectrum' of fringe patterns each of slightly different spacing. If all the fringe patterns are in phase in the centre, then the fringes will increase in size as the wavelength decreases and the summed intensity will show three to four fringes of varying colour. Young describes this very elegantly in his discussion of two slit interference. Some fine examples of white light fringes can be seen here. Since white light fringes are obtained only when the two waves have travelled equal distances from the light source, they can be very useful in interferometry, as they allow the zero path difference fringe to be identified.

Optical Arrangements

To generate interference fringes, light from the source has to be divided into two waves which have then to be re-combined. Traditionally, interferometers have been classified as either amplitude-division or wavefront-division systems.

In an amplitude-division system, a beam splitter is used to divide the light into two beams travelling in different directions, which are then superimposed to produce the interference pattern. The Michelson interferometer and the Mach-Zehnder interferometer are examples of amplitude-division systems.

In wavefront-division systems, the wave is divided in space—examples are Young's double slit interferometer and Lloyd's mirror.

Interference can also be seen in everyday phenomena such as iridescence and structural coloration. For example, the colours seen in a soap bubble arise from interference of light reflecting off the front and back surfaces of the thin soap film. Depending on the thickness of the film, different colours interfere constructively and destructively.

Applications

Interferometry has played an important role in the advancement of physics, and also has a wide range of applications in physical and engineering measurement.

Thomas Young's double slit interferometer in 1803 demonstrated interference fringes when two small holes were illuminated by light from another small hole which was illuminated by sunlight. Young was able to estimate the wavelength of different colours in the spectrum from the spacing of the fringes. The experiment played a major role in the general acceptance of the wave theory of light. In quantum mechanics, this experiment is considered to demonstrate the inseparability of the wave and particle natures of light and other quantum particles (wave–particle duality). Richard Feynman was fond of saying that all of quantum mechanics can be gleaned from carefully thinking through the implications of this single experiment.

The results of the Michelson–Morley experiment are generally considered to be the first strong evidence against the theory of a luminiferous aether and in favor of special relativity.

Interferometry has been used in defining and calibrating length standards. When the metre was defined as the distance between two marks on a platinum-iridium bar, Michelson and Benoît used interferometry to measure the wavelength of the red cadmium line in the new standard, and also showed that it could be used as a length standard. Sixty years later, in 1960, the metre in the new SI system was defined to be equal to 1,650,763.73 wavelengths of the orange-red emission line in the electromagnetic spectrum of the krypton-86 atom in a vacuum. This definition was replaced in 1983 by defining the metre as the distance travelled by light in vacuum during a specific time interval. Interferometry is still fundamental in establishing the calibration chain in length measurement.

Interferometry is used in the calibration of slip gauges (called gauge blocks in the US) and in coordinate-measuring machines. It is also used in the testing of optical components.

Radio Interferometry

The Very Large Array, an interferometric array formed from many smaller telescopes, like many larger radio telescopes.

In 1946, a technique called astronomical interferometry was developed. Astronomical radio interferometers usually consist either of arrays of parabolic dishes or two-dimensional arrays of omni-directional antennas. All of the telescopes in the array are widely separated and are usually connected together using coaxial cable, waveguide, optical fiber, or other type of transmission line. Interferometry increases the total signal collected, but its primary purpose is to vastly increase the resolution through a process called Aperture synthesis. This technique works by superposing (interfering) the signal waves from the different telescopes on the principle that waves that coincide with the same phase will add to each other while two waves that have opposite phases will cancel each other out. This creates a combined telescope that is equivalent in resolution (though not in sensitivity) to a single antenna whose diameter is equal to the spacing of the antennas furthest apart in the array.

Acoustic Interferometry

An acoustic interferometer is an instrument for measuring the physical characteristics of sound wave in a gas or liquid. It may be used to measure velocity, wavelength, absorption, or impedance. A vibrating crystal creates the ultrasonic waves that are radiated into the medium. The waves strike a reflector placed parallel to the crystal. The waves are then reflected back to the source and measured.

Quantum Interference

If a system is in state ψ, its wavefunction is described in Dirac or bra–ket notation as:

$$|\psi\rangle = \sum_i |i\rangle\langle i|\psi\rangle = \sum_i |i\rangle\psi_i$$

where the $|i\rangle$ s specify the different quantum "alternatives" available (technically, they form an eigenvector basis) and the ψ_i are the probability amplitude coefficients, which are complex numbers.

The probability of observing the system making a transition or quantum leap from state ψ to a new state φ is the square of the modulus of the scalar or inner product of the two states:

$$\text{prob}(\psi \Rightarrow \varphi) = |\langle\psi|\varphi\rangle|^2 = |\sum_i \psi_i^* \varphi_i|^2$$
$$= \sum_{ij} \psi_i^* \psi_j \varphi_j^* \varphi_i = \sum_i |\psi_i|^2 |\varphi_i|^2 + \sum_{ij;i\neq j} \psi_i^* \psi_j \varphi_j^* \varphi_i$$

where $\psi_i = \langle i|\psi\rangle$ (as defined above) and similarly $\varphi_i = \langle i|\varphi\rangle$ are the coefficients of the final state of the system. * is the complex conjugate so that $\psi_i^* = \langle\psi|i\rangle$, etc.

Now let's consider the situation classically and imagine that the system transited from $|\psi\rangle$ to $|\varphi\rangle$ via an intermediate state $|i\rangle$. Then we would *classically* expect the probability of the two-step transition to be the sum of all the possible intermediate steps. So we would have

$$\text{prob}(\psi \Rightarrow \varphi) = \sum_i \text{prob}(\psi \Rightarrow i \Rightarrow \varphi)$$
$$= \sum_i |\langle\psi|i\rangle|^2 |\langle i|\varphi\rangle|^2 = \sum_i |\psi_i|^2 |\varphi_i|^2,$$

The classical and quantum derivations for the transition probability differ by the presence, in the quantum case, of the extra terms $\sum_{ij;i\neq j} \psi_i^* \psi_j \varphi_j^* \varphi_i$; these extra quantum terms represent *interference* between the different $i \neq j$ intermediate "alternatives". These are consequently known as the quantum interference terms, or cross terms. This is a purely quantum effect and is a consequence of the non-additivity of the probabilities of quantum alternatives.

The interference terms vanish, via the mechanism of quantum decoherence, if the intermediate state $|i\rangle$ is measured or coupled with the environment.

Powder Diffraction

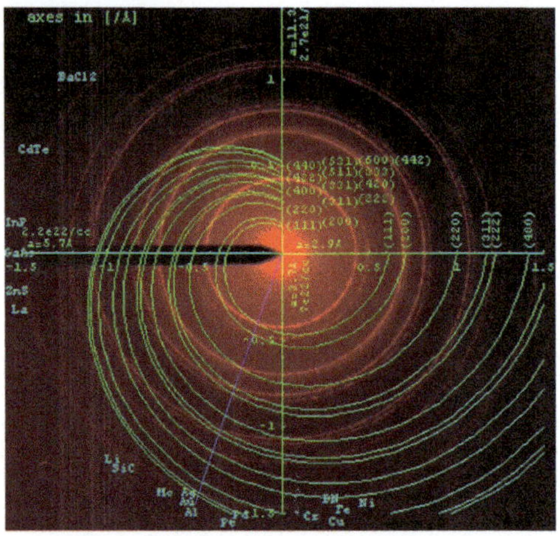

Electron powder pattern (red) of an Al film with an fcc spiral overlay (green) and a line of intersections (blue) that determines lattice parameter.

Powder diffraction is a scientific technique using X-ray, neutron, or electron diffraction on powder or microcrystalline samples for structural characterization of materials. An instrument dedicated to performing such powder measurements is called a powder diffractometer.

Powder diffraction stands in contrast to single crystal diffraction techniques, which work best with a single, well-ordered crystal.

Explanation

A diffractometer produces waves at a known frequency, which is determined by their source. The source is often x-rays, because they are the only kind of light with the correct frequency for inter-atomic-scale diffraction. However, electrons and neutrons are also common sources, with their frequency determined by their de Broglie wavelength. When these waves reach the sample, the atoms of the sample act just like a diffraction grating, producing bright spots at particular angles. By measuring the angle where these bright spots occur, the spacing of the diffraction grating can be determined by Bragg's law. Because the sample itself is the diffraction grating, this spacing is the atomic spacing.

The distinction between powder and single crystal diffraction is the degree of texturing in the sample. Single crystals have maximal texturing, and are said to be anisotropic. In contrast, in powder diffraction, every possible crystalline orientation is represented equally in a powdered sample, the isotropic case. PXRD operates under the assumption that the sample is randomly arranged. Therefore, a statistically significant number of each plane of the crystal structure will be in the proper orientation to diffract the X-rays. Therefore, each plane will be represented in the signal. In practice, it is sometimes necessary to rotate the sample orientation to eliminate the effects of texturing and achieve true randomness.

Mathematically, crystals can be described by a Bravais lattice with some regularity in the spacing between atoms. Because of this regularity, we can describe this structure in a different way using the reciprocal lattice, which is related to the original structure by a Fourier transform. This three-dimensional space can be described with reciprocal axes x*, y*, and z* or alternatively in spherical coordinates q, φ*, and χ*. In powder diffraction, intensity is homogeneous over φ* and χ*, and only q remains as an important measurable quantity. This is because orientational averaging causes the three-dimensional reciprocal space that is studied in single crystal diffraction to be projected onto a single dimension.

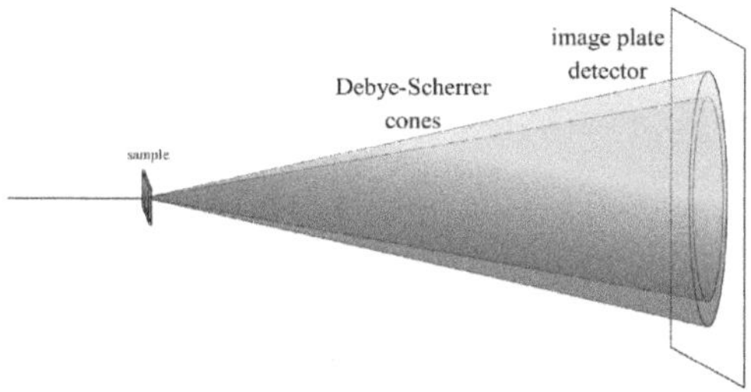

Two-dimensional powder diffraction setup with flat plate detector.

When the scattered radiation is collected on a flat plate detector, the rotational averaging leads to smooth diffraction rings around the beam axis, rather than the discrete Laue spots observed in single crystal diffraction. The angle between the beam axis and the ring is called the *scattering angle* and in X-ray crystallography always denoted as 2θ (in scattering of *visible* light the convention is usually to call it θ). In accordance with Bragg's law, each ring corresponds to a particular reciprocal lattice vector G in the sample crystal. This leads to the definition of the scattering vector as:

$$|G| = q = 2k\sin(\theta) = \frac{4\pi}{\lambda}\sin(\theta).$$

In this equation, G is the reciprocal lattice vector, q is the length of the reciprocal lattice vector, k is the momentum transfer vector, θ is half of the scattering angle, and λ is the wavelength of the radiation. Powder diffraction data are usually presented as a diffractogram in which the diffracted intensity, I, is shown as a function either of the scattering angle 2θ or as a function of the scattering vector length q. The latter variable has the advantage that the diffractogram no longer depends on the value of the wavelength λ. The advent of synchrotron sources has widened the choice of wavelength considerably. To facilitate comparability of data obtained with different wavelengths the use of q is therefore recommended and gaining acceptability.

Uses

Relative to other methods of analysis, powder diffraction allows for rapid, non-destructive analysis of multi-component mixtures without the need for extensive sample preparation. This gives laboratories around the world the ability to quickly analyze unknown materials and perform materials characterization in such fields as metallurgy, mineralogy, forensic science, archeology, condensed matter physics, and the biological and pharmaceutical sciences. Identification is performed by comparison of the diffraction pattern to a known standard or to a database such as the International Centre for Diffraction Data's Powder Diffraction File (PDF) or the Cambridge Structural Database (CSD). Advances in hardware and software, particularly improved optics and fast detectors, have dramatically improved the analytical capability of the technique, especially relative to the speed of the analysis. The fundamental physics upon which the technique is based provides high precision and accuracy in the measurement of interplanar spacings, sometimes to fractions of an Ångström, resulting in authoritative identification frequently used in patents, criminal cases and other areas of law enforcement. The ability to analyze multiphase materials also allows analysis of how materials interact in a particular matrix such as a pharmaceutical tablet, a circuit board, a mechanical weld, a geologic core sampling, cement and concrete, or a pigment found in an historic painting. The method has been historically used for the identification and classification of minerals, but it can be used for any materials, even amorphous ones, so long as a suitable reference pattern is known or can be constructed.

Phase Identification

The most widespread use of powder diffraction is in the identification and characterization of crystalline solids, each of which produces a distinctive diffraction pattern. Both the positions (corresponding to lattice spacings) and the relative intensity of the lines in a diffraction pattern are indicative of a particular phase and material, providing a "fingerprint" for comparison. A multiphase mixture, e.g. a soil sample, will show more than one pattern superposed, allowing for determination of the relative concentrations of phases in the mixture.

J.D. Hanawalt, an analytical chemist who worked for Dow Chemical in the 1930s, was the first to realize the analytical potential of creating a database. Today it is represented by the Powder Diffraction File (PDF) of the International Centre for Diffraction Data (formerly Joint Committee for Powder Diffraction Studies). This has been made searchable by computer through the work of global software developers and equipment manufacturers. There are now over 550,000 reference materials in the 2006 Powder Diffraction File Databases, and these databases are interfaced to a wide variety of diffraction analysis software and distributed globally. The Powder Diffraction File contains many subfiles, such as minerals, metals and alloys, pharmaceuticals, forensics, excipients, superconductors, semiconductors, etc., with large collections of organic, organometallic and inorganic reference materials.

Crystallinity

In contrast to a crystalline pattern consisting of a series of sharp peaks, amorphous materials (liquids, glasses etc.) produce a broad background signal. Many polymers show semicrystalline behavior, *i.e.* part of the material forms an ordered crystallite by folding of the molecule. A single polymer molecule may well be folded into two different, adjacent crystallites and thus form a tie between the two. The tie part is prevented from crystallizing. The result is that the crystallinity will never reach 100%. Powder XRD can be used to determine the crystallinity by comparing the integrated intensity of the background pattern to that of the sharp peaks. Values obtained from powder XRD are typically comparable but not quite identical to those obtained from other methods such as DSC.

Lattice Parameters

The position of a diffraction peak is 'independent' of the atomic positions within the cell and entirely determined by the size and shape of the unit cell of the crystalline phase. Each peak represents a certain lattice plane and can therefore be characterized by a Miller index. If the symmetry is high, e.g.: cubic or hexagonal it is usually not too hard to identify the index of each peak, even for an unknown phase. This is particularly important in solid-state chemistry, where one is interested in finding and identifying new materials. Once a pattern has been indexed, this characterizes the reaction product and identifies it as a new solid phase. Indexing programs exist to deal with the harder cases, but if the unit cell is very large and the symmetry low (triclinic) success is not always guaranteed.

Expansion Tensors, Bulk Modulus

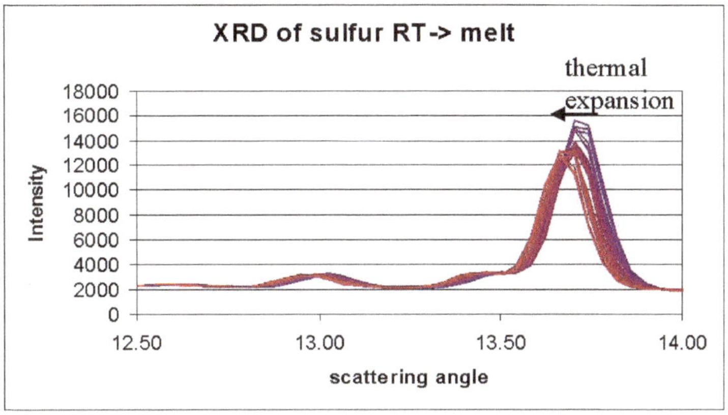

Thermal expansion of a sulfur powder

Cell parameters are somewhat temperature and pressure dependent. Powder diffraction can be combined with *in situ* temperature and pressure control. As these thermodynamic variables are changed, the observed diffraction peaks will migrate continuously to indicate higher or lower lattice spacings as the unit cell distorts. This allows for measurement of such quantities as the thermal expansion tensor and the isothermal bulk modulus, as well determination of the full equation of state of the material.

Phase Transitions

At some critical set of conditions, for example 0 °C for water at 1 atm, a new arrangement of atoms or molecules may become stable, leading to a phase transition. At this point new diffraction peaks will appear or old ones disappear according to the symmetry of the new phase. If the material melts to an isotropic liquid, all sharp lines will disappear and be replaced by a broad amorphous pattern. If the transition produces another crystalline phase, one set of lines will suddenly be replaced by another set. In some cases however lines will split or coalesce, e.g. if the material undergoes a continuous, second order phase transition. In such cases the symmetry may change because the existing structure is *distorted* rather than replaced by a completely different one. For example, the diffraction peaks for the lattice planes (100) and (001) can be found at two different values of q for a tetragonal phase, but if the symmetry becomes cubic the two peaks will come to coincide.

Crystal Structure Refinement and Determination

Crystal structure determination from powder diffraction data is extremely challenging due to the overlap of reflections in a powder experiment. A number of different methods exist for structural determination, such as simulated annealing and charge flipping. The crystal structures of known materials can be refined, i.e. as a function of temperature or pressure, using the Rietveld method. The Rietveld method is a so-called full pattern analysis technique. A crystal structure, together with instrumental and microstructural information, is used to generate a theoretical diffraction pattern that can be compared to the observed data. A least squares procedure is then used to minimize the difference between the calculated pattern and each point of the observed pattern by adjusting model parameters. Techniques to determine unknown structures from powder data do exist, but are somewhat specialized. A number of programs that can be used in structure determination are TOPAS, Fox, DASH, GSAS, EXPO2004, and a few others.

Size and Strain Broadening

There are many factors that determine the width B of a diffraction peak. These include:

1. instrumental factors
2. the presence of defects to the perfect lattice
3. differences in strain in different grains
4. the size of the crystallites

It is often possible to separate the effects of size and strain. Where size broadening is independent of q (K=1/d), strain broadening increases with increasing q-values. In most cases there will be both size and strain broadening. It is possible to separate these by combining the two equations in what is known as the Hall–Williamson method:

$$B \cdot \cos(\theta) = \frac{k\lambda}{D} + \eta \cdot \sin(\theta),$$

Thus, when we plot $B \cdot \cos(\theta)$ vs. $\sin(\theta)$ we get a straight line with slope η and intercept $\frac{k\lambda}{D}$. .

The expression is a combination of the Scherrer equation for size broadening and the Stokes and Wilson expression for strain broadening. The value of η is the strain in the crystallites, the value of D represents the size of the crystallites. The constant *k* is typically close to unity and ranges from 0.8 to 1.39.

Comparison of X-ray and Neutron Scattering

X-ray photons scatter by interaction with the electron cloud of the material, neutrons are scattered by the nuclei. This means that, in the presence of heavy atoms with many electrons, it may be difficult to detect light atoms by X-ray diffraction. In contrast, the neutron scattering lengths of most atoms are approximately equal in magnitude. Neutron diffraction techniques may therefore be used to detect light elements such as oxygen or hydrogen in combination with heavy atoms. The neutron diffraction technique therefore has obvious applications to problems such as determining oxygen displacements in materials like high temperature superconductors and ferroelectrics, or to hydrogen bonding in biological systems.

A further complication in the case of neutron scattering from hydrogenous materials is the strong incoherent scattering of hydrogen (80.27(6) barn). This leads to a very high background in neutron diffraction experiments, and may make structural investigations impossible. A common solution is deuteration, i.e., replacing the 1-H atoms in the sample with deuterium (2-H). The incoherent scattering length of deuterium is much smaller (2.05(3) barn) making structural investigations significantly easier. However, in some systems, replacing hydrogen with deuterium may alter the structural and dynamic properties of interest.

As neutrons also have a magnetic moment, they are additionally scattered by any magnetic moments in a sample. In the case of long range magnetic order, this leads to the appearance of new Bragg reflections. In most simple cases, powder diffraction may be used to determine the size of the moments and their spatial orientation.

Aperiodically Arranged Clusters

Predicting the scattered intensity in powder diffraction patterns from gases, liquids, and randomly distributed nano-clusters in the solid state is (to first order) done rather elegantly with the Debye scattering equation:

$$I(q) = \sum_{i=1}^{N}\sum_{j=1}^{N} f_i(q) f_j(q) \frac{\sin(q r_{ij})}{q r_{ij}},$$

where the magnitude of the scattering vector q is in reciprocal lattice distance units, N is the number of atoms, $f_i(q)$ is the atomic scattering factor for atom i and scattering vector q, while r_{ij} is the distance between atom i and atom j. One can also use this to predict the effect of nano-crystallite shape on detected diffraction peaks, even if in some directions the cluster is only one atom thick.

Devices

Cameras

The simplest cameras for X-ray powder diffraction consist of a small capillary and either a flat plate detector (originally a piece of X-ray film, now more and more a flat-plate detector or a CCD-camera) or a cylindrical one (originally a piece of film in a cookie-jar, but increasingly bent position sensitive detectors are used). The two types of cameras are known as the Laue and the Debye–Scherrer camera.

In order to ensure complete powder averaging, the capillary is usually spun around its axis.

For neutron diffraction vanadium cylinders are used as sample holders. Vanadium has a negligible absorption and coherent scattering cross section for neutrons and is hence nearly invisible in a powder diffraction experiment. Vanadium does however have a considerable incoherent scattering cross section which may cause problems for more sensitive techniques such as neutron inelastic scattering.

A later development in X-ray cameras is the Guinier camera. It is built around a *focusing* bent crystal monochromator. The sample is usually placed in the focusing beam, e.g. as a dusting on a piece of sticky tape. A cylindrical piece of film (or electronic multichannel detector) is put on the focusing circle, but the incident beam prevented from reaching the detector to prevent damage from its high intensity.

Diffractometers

Diffractometers can be operated both in transmission and in reflection configurations. The reflection one is more common. The powder sample is filled in a small disc-like container and its surface carefully flattened. The disc is put on one axis of the diffractometer and tilted by an angle θ while a detector (scintillation counter) rotates around it on an arm at twice this angle. This configuration is known under the name Bragg–Brentano theta-2 theta.

Another configuration is the Bragg–Brentano theta-theta configuration in which the sample is stationary while the X-ray tube and the detector are rotated around it. The angle formed between the tube and the detector is 2theta. This configuration is most convenient for loose powders.

Position-sensitive detectors (PSD) and area detectors, which allow collection from multiple angles at once, are becoming more popular on currently supplied instrumentation.

Neutron Diffraction

Sources that produce a neutron beam of suitable intensity and speed for diffraction are only available at a small number of research reactors and spallation sources in the world. Angle dispersive (fixed wavelength) instruments typically have a battery of individual detectors arranged in a cylindrical fashion around the sample holder, and can therefore collect scattered intensity simultaneously on a large 2θ range. Time of flight instruments normally have a small range of banks at different scattering angles which collect data at varying resolutions.

X-ray Tubes

Laboratory X-ray diffraction equipment relies on the use of an X-ray tube, which is used to produce the X-rays. The most commonly used laboratory X-ray tube uses a copper anode, but cobalt and molybdenum are also popular. The wavelength in nm varies for each source. The table below shows these wavelengths, determined by Bearden and quoted in the International Tables for X-ray Crystallography (all values in nm):

Element	**Kα (weight average)**	**Kα2 (strong)**	**Kα1 (very strong)**	**Kβ (weak)**
Cr	0.229100	0.229361	0.228970	0.208487
Fe	0.193736	0.193998	0.193604	0.175661
Co	0.179026	0.179285	0.178897	0.162079
Cu	0.154184	0.154439	0.154056	0.139222
Mo	0.071073	0.071359	0.070930	0.063229

According to the last re-examination of Holzer et al. (1997), these values are respectively:

Element	Kα2	Kα1	Kβ
Cr	0.2293663	0.2289760	0.2084920
Co	0.1792900	0.1789010	0.1620830
Cu	0.1544426	0.1540598	0.1392250
Mo	0.0713609	0.0709319	0.0632305

Other Sources

In-house applications of X-ray diffraction has always been limited to the relatively few wavelengths shown in the table above. The available choice was much needed because the combination of certain wavelengths and certain elements present in a sample can lead to strong fluorescence which increases the background in the diffraction pattern. A notorious example is the presence of iron in a sample when using copper radiation. In general elements just below the anode element in the period system need to be avoided.

Another limitation is that the intensity of traditional generators is relatively low, requiring lengthy exposure times and precluding any time dependent measurement. The advent of synchrotron sources has drastically changed this picture and caused powder diffraction methods to enter a whole new phase of development. Not only is there a much wider choice of wavelengths available, the high brilliance of the synchrotron radiation makes it possible to observe changes in the pattern during chemical reactions, temperature ramps, changes in pressure and the like.

The tunability of the wavelength also makes it possible to observe anomalous scattering effects when the wavelength is chosen close to the absorption edge of one of the elements of the sample.

Neutron diffraction has never been an in house technique because it requires the availability of an intense neutron beam only available at a nuclear reactor or spallation source. Typically the available neutron flux, and the weak interaction between neutrons and matter, require relative large samples.

Advantages and Disadvantages

Although it is possible to solve crystal structures from powder X-ray data alone, its single crystal analogue is a far more powerful technique for structure determination. This is directly related to the fact that much information is lost by the collapse of the 3D space onto a 1D axis. Nevertheless powder X-ray diffraction is a powerful and useful technique in its own right. It is mostly used to characterize and identify *phases*, and to refine details of an already known structure, rather than solving unknown structures.

Advantages of the technique are:

- simplicity of sample preparation
- rapidity of measurement
- the ability to analyze mixed phases, e.g. soil samples
- "in situ" structure determination

By contrast growth and mounting of large single crystals is notoriously difficult. In fact there are many materials for which, despite many attempts, it has not proven possible to obtain single crystals. Many materials are readily available with sufficient microcrystallinity for powder diffraction, or samples may be easily ground from larger crystals. In the field of solid-state chemistry that often aims at synthesizing *new* materials, single crystals thereof are typically not immediately available. Powder diffraction is therefore one of the most powerful methods to identify and characterize new materials in this field.

Particularly for neutron diffraction, which requires larger samples than X-ray diffraction due to a relatively weak scattering cross section, the ability to use large samples can be critical, although newer and more brilliant neutron sources are being built that may change this picture.

Since all possible crystal orientations are measured simultaneously, collection times can be quite short even for small and weakly scattering samples. This is not merely convenient, but can be essential for samples which are unstable either inherently or under X-ray or neutron bombardment, or for time-resolved studies. For the latter it is desirable to have a strong radiation source. The advent of synchrotron radiation and modern neutron sources has therefore done much to revitalize the powder diffraction field because it is now possible to study temperature dependent changes, reaction kinetics and so forth by means of time-resolved powder diffraction.

- The identification of crystal structure of unknown materials has been performed usually with a powder XRD. Powders, containing aggregation of fine grains with a single crystal structure, but oriented randomly, when exposed to X-rays, the diffraction intensity is assumed to be the sum of the X-rays reflected from all the fine grains. Hence, we can observe the diffraction peak attributed to Miller indices of the powders.
- For known materials, there is a database maintained by the International Center for Diffraction Data (ICDD) or Joint Committee on Powder Diffraction Standards (JCPDS) for the identification of inorganic and organic materials with a crystal structure.
- One can identify the unknown materials by comparing their diffraction patterns and the database. In addition, XRD patterns can be used for (i) identification of crystalline phases and their degree of crystallization, (ii) estimating crystallite size and (iii) residual strain.

Information Obtainable from a Diffraction Pattern

Figure depicts the schematic of XRD pattern and information obtainable.

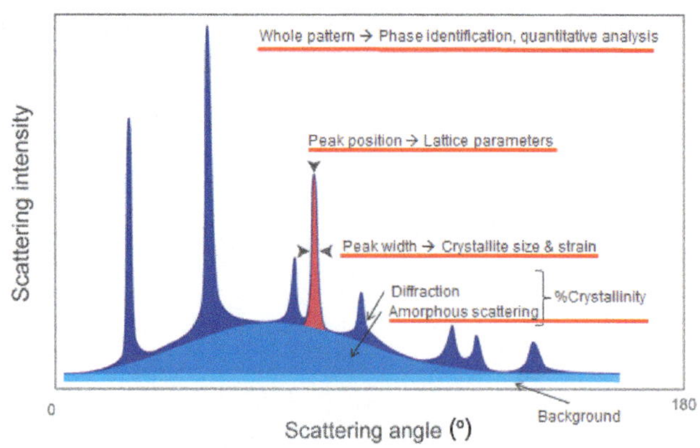

Schematic of XRD pattern.

XRD patterns can be used to determine

- Phase Composition of a sample:
  - Quantitative Phase Analysis can determine the relative amounts of phases in a mixture by referencing the relative peak intensities
- Unit cell lattice parameters and Bravais lattice symmetry
  - Index peak positions
  - Lattice parameters can vary and thereby giving the information about, alloying, doping, solid solutions, strains, etc.
- Residual Strain (macrostrain)
- Crystal Structure – By Rietveld refinement of the entire diffraction pattern
- Epitaxy/Texture/Orientation
- Crystallite Size and Microstrain
  - Indicated by peak broadening
  - Other defects (stacking faults, etc.) can be measured by analysis of peak shapes and peak width
- The above factors can be also determined as a function of time, temperature, and gas environment in some of the advanced XRD having in-situ heating and controlled atmosphere facilities.

Phase Identification

Figure displays the typical XRD pattern of a unknown sample, the process of comparing with the database and identifying the sample structure after matching with the database. This process is almost similar to the one done for finger print search.

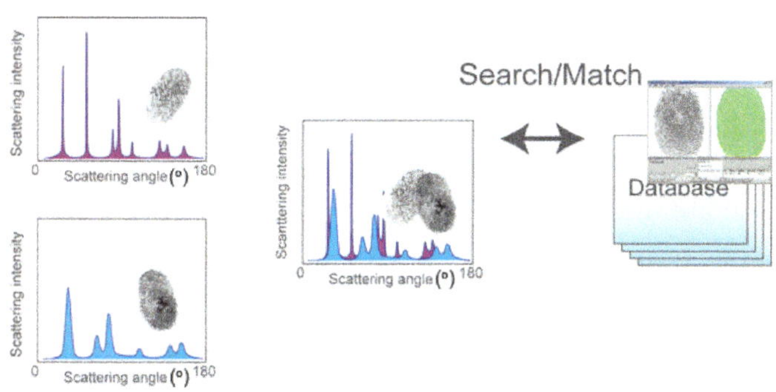

Schamtic of identifying the sample structure using database search.

It may be noted that the diffraction pattern for every phase is as unique as our fingerprint, i.e.,

- Phases with the same chemical composition can have drastically different diffraction patterns.

- Use the position and relative intensity of a series of peaks to match experimental data to the reference patterns in the database

X-ray Diffraction Patterns for Different Structures

Single Crystal

A single crystal specimen in a Bragg-Brentano diffractometer would produce only one family of peaks in the diffraction pattern as shown in figure.

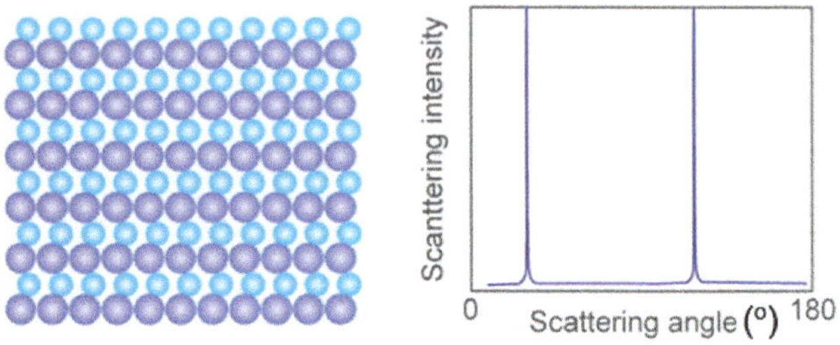

Typical XRD pattern of single crystal specimen.

Powders / Polycrystals

Powders / polycrystalline samples contain thousands of crystallites with each one having a single crystal structure. However, each of the crystals are oriented randomly. Therefore, all possible diffraction peaks should be observed.

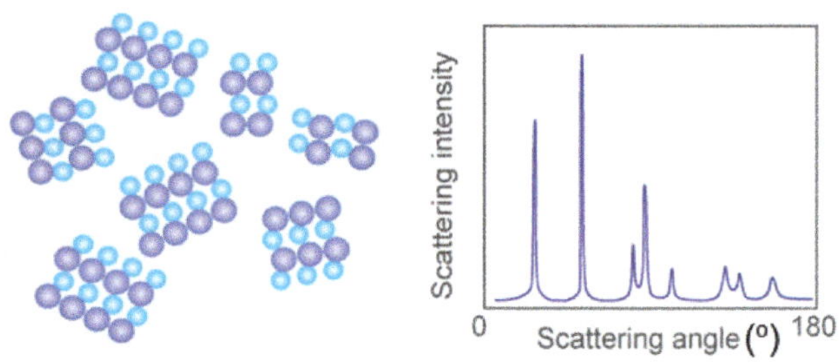

Typical XRD pattern of polycrystal specimen.

X-ray diffraction Patterns for different structures:

Amorphous/ Liquid:

Amorphous/Liquid samples contain a large number of atoms without forming any major crystalline phases. Hence, there is no possibility to get a clear diffraction peak in such a specimen. As a result, the diffraction from such structure provides a broad peak as shown in figure.

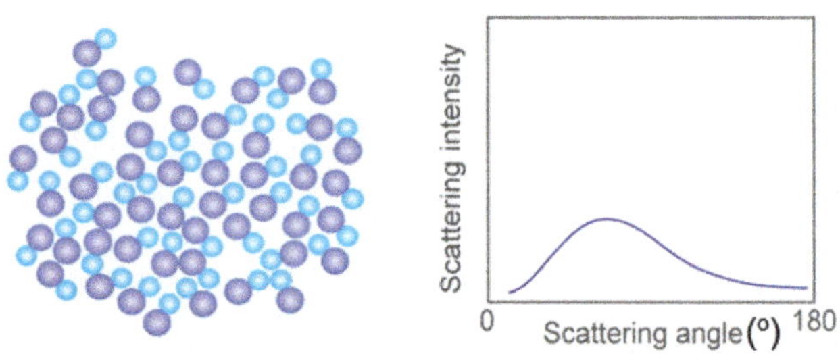

Typical XRD pattern of amorphous or liquid specimen.

Gas molecules

Gas molecules move randomly within a given volume resulting no significant diffraction. Hence, as demonstrated in figure, the intensity of X-ray drops largely with increasing angle and attains zero.

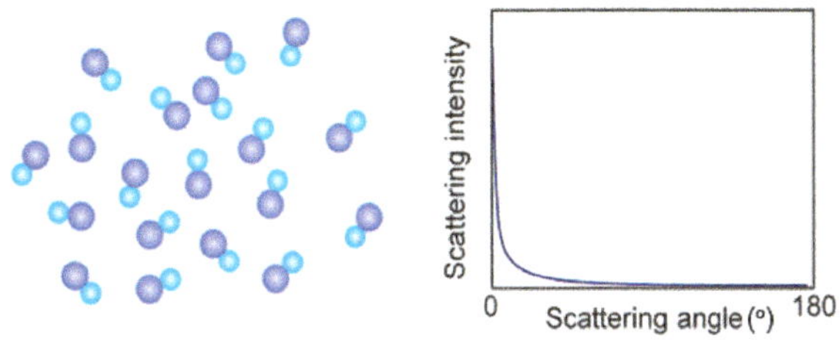

Typical XRD pattern of gas molecules.

The Structural Make up of Solids

- Small disordered cells are referred to be as amorphous, resulting in a broad diffraction pattern.
- Crystallites show a sufficient extent of repeated unit cells as displayed in figure to display diffraction.
- Random orientation of crystallites produces the typical diffraction pattern.
- Preferred orientation of crystallites will produce a significant distortion of diffraction intensities from the ideal.

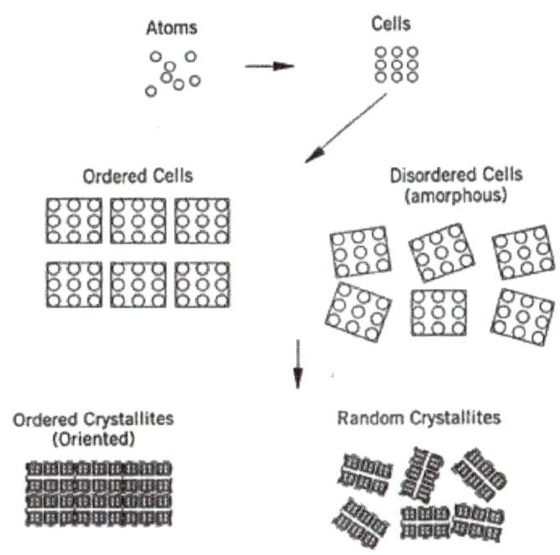

Schematic of cells to form solids.

Crystallite

A crystallite is a small or even microscopic crystal which forms, for example, during the cooling of many materials. The orientation of crystallites can be random with no preferred direction, called random texture, or directed, possibly due to growth and processing conditions. Fiber texture is an example of the latter. Crystallites are also referred to as grains. The areas where crystallite grains meet are known as grain boundaries. Polycrystalline or multicrystalline materials, or polycrystals are solids that are composed of many crystallites of varying size and orientation.

Polycrystalline structures composed of crystallites. *Clockwise from top left:*
a) malleable iron
b) electrical steel without coating
c) solar cells made of multicrystalline silicon
d) galvanized surface of zinc
e) micrograph of acid etched metal highlighting grain boundaries

Most inorganic solids are polycrystalline, including all common metals, many ceramics, rocks and ice. The extent to which a solid is crystalline (crystallinity) has important effects on its physical properties. Sulfur, while usually polycrystalline, may also occur in other allotropic forms with completely different properties. Although crystallites are referred to as grains, *powder grains* are different, as they can be composed of smaller polycrystalline grains themselves.

While the structure of a (monocrystalline) crystal is highly ordered and its lattice is continuous and unbroken. amorphous materials, such as glass and many polymers, are non-crystalline and do not display any structures as their constituents are not arranged in an ordered manner. Polycrystalline structures and paracrystalline phases are in between these two extremes.

Details

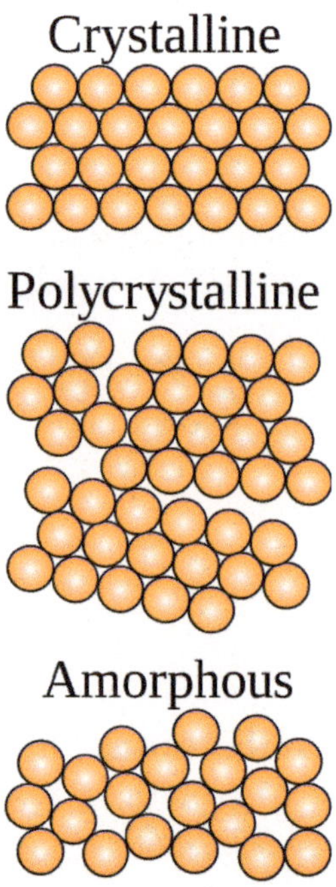

Different degrees of ordered structures: a monocrystalline crystal, polycrystalline structure, and amorphous or non-crystalline solid.

Crystal size is usually measured from X-ray diffraction patterns and grain size by other experimental techniques like transmission electron microscopy. Solid objects that are large enough to see and handle are rarely composed of a single crystal, except for a few cases (gems, silicon single crystals for the electronics industry, certain types of fiber, single crystals of a nickel-based superalloy for turbojet engines, and some ice crystals which can exceed 0.5 meters in diameter). Most materials are *polycrystalline*; they are made of a large number of single crystals — crystallites — held together by thin layers of amorphous solid. The crystallite size can vary from a few nanometers to several millimeters.

Bronze bell with large crystallites on the inside.

If the individual crystallites are oriented completely at random, a large enough volume of polycrystalline material will be approximately isotropic. This property helps the simplifying assumptions of continuum mechanics to apply to real-world solids. However, most manufactured materials have some alignment to their crystallites, resulting in *texture* that must be taken into account for accurate predictions of their behavior and characteristics. When the crystallites are mostly ordered with just some random spread of orientations, one has a mosaic crystal.

Material fractures can be intergranular fracture or a transgranular fracture. There is an ambiguity with powder grains: a powder grain can be made of several crystallites. Thus, the (powder) "grain size" found by laser granulometry can be different from the "grain size" (or, rather, crystallite size) found by X-ray diffraction (e.g. Scherrer method), by optical microscopy under polarised light, or by scanning electron microscopy (backscattered electrons).

Coarse grained rocks are formed very slowly, while fine grained rocks are formed quickly, on geological time scales. If a rock forms very quickly, such as the solidification of lava ejected from a volcano, there may be no crystals at all. This is how obsidian forms.

Grain Boundaries

Grain boundaries are interfaces where crystals of different orientations meet. A grain boundary is a single-phase interface, with crystals on each side of the boundary being identical except in orientation. The term "crystallite boundary" is sometimes, though rarely, used. Grain boundary areas contain those atoms that have been perturbed from their original lattice sites, dislocations, and impurities that have migrated to the lower energy grain boundary.

Treating a grain boundary geometrically as an interface of a single crystal cut into two parts, one of which is rotated, we see that there are five variables required to define a grain boundary. The first two numbers come from the unit vector that specifies a rotation axis. The third number designates the angle of rotation of the grain. The final two numbers specify the plane of the grain boundary (or a unit vector that is normal to this plane).

Grain boundaries disrupt the motion of dislocations through a material. Dislocation propagation is impeded because of the stress field of the grain boundary defect region and the lack of slip planes and slip directions and overall alignment across the boundaries. Reducing grain size is therefore a common way to improve strength, often without any sacrifice in toughness because the smaller grains create more obstacles per unit area of slip plane. This crystallite size-strength relationship is given by the Hall-Petch relationship. The high interfacial energy and relatively weak bonding in grain boundaries makes them preferred sites for the onset of corrosion and for the precipitation of new phases from the solid.

Grain boundary migration plays an important role in many of the mechanisms of creep. Grain boundary migration occurs when a shear stress acts on the grain boundary plane and causes the grains to slide. This means that fine-grained materials actually have a poor resistance to creep relative to coarser grains, especially at high temperatures, because smaller grains contain more atoms in grain boundary sites. Grain boundaries also cause deformation in that they are sources and sinks of point defects. Voids in a material tend to gather in a grain boundary, and if this happens to a critical extent, the material could fracture.

During grain boundary migration, the rate determining step depends on the angle between two adjacent grains. In a small angle dislocation boundary, the migration rate depends on vacancy diffusion between dislocations. In a high angle dislocation boundary, this depends on the atom transport by single atom jumps from the shrinking to the growing grains.

Grain boundaries are generally only a few nanometers wide. In common materials, crystallites are large enough that grain boundaries account for a small fraction of the material. However, very small grain sizes are achievable. In nanocrystalline solids, grain boundaries become a significant volume fraction of the material, with profound effects on such properties as diffusion and plasticity. In the limit of small crystallites, as the volume fraction of grain boundaries approaches 100%, the material ceases to have any crystalline character, and thus becomes an amorphous solid.

Grain boundaries are also present in magnetic domains in magnetic materials. A computer hard disk, for example, is made of a hard ferromagnetic material that contains regions of atoms whose magnetic moments can be realigned by an inductive head. The magnetization varies from region to region, and the misalignment between these regions forms boundaries that are key to data storage. The inductive head measures the orientation of the magnetic moments of these domain regions and reads out either a "1" or "0". These bits are the data being read. Grain size is important in this technology because it limits the number of bits that can fit on one hard disk. The smaller the grain sizes, the more data that can be stored.

Because of the dangers of grain boundaries in certain materials such as superalloy turbine blades, great technological leaps were made to minimize as much as possible the effect of grain boundaries in the blades. The result was directional solidification processing in which grain boundaries were eliminated by producing columnar grain structures aligned parallel to the axis of the blade, since this is usually the direction of maximum tensile stress felt by a blade during its rotation in an airplane. The resulting turbine blades consisted of a single grain, improving reliability.

Generally, polycrystals cannot be superheated; they will melt promptly once they are brought to a high enough temperature. This is because grain boundaries are amorphous, and serve as nucleation points for the liquid phase. By contrast, if no solid nucleus is present as a liquid cools, it tends to become supercooled. Since this is undesirable for mechanical materials, alloy designers often take steps against it.

Calculation of Average Crystallite Size

Let us consider a large sized crystal as shown in figure. This would exhibit a XRD peak occurring only at a particular angle $2\theta_B$.

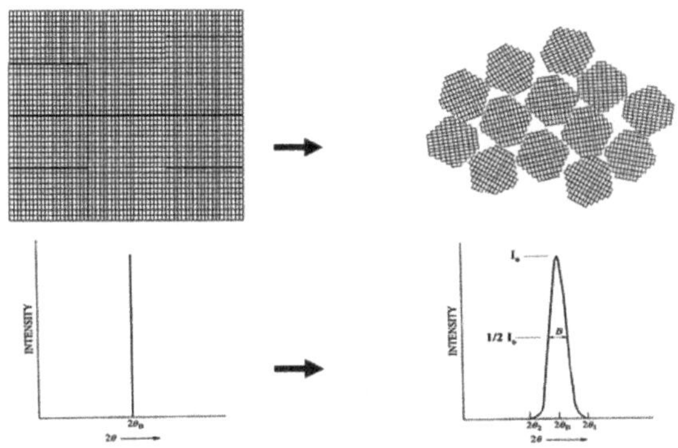

Schematic of large sized crystal, crystals with reduced size and the corresponding XRD patterns.

On the other hand, the width of the diffraction peak increases as the size of the crystal decreases. This allows one to calculate the average crystallite size from the XRD peak broadening.

Consider a crystal having thickness t as shown in figure measured in a direction perpendicular (⊥) to a set of reflecting planes.

Let there be (m +1) planes in this set.

The Bragg angle θ as a variable and call θ_B satisfies exactly the Bragg law, resulting in the diffraction condition

$$\lambda = 2d \sin\theta_B$$

Intensity of the beams diffracted at and above $2\theta_1$, and at and below $2\theta_2$ is zero.

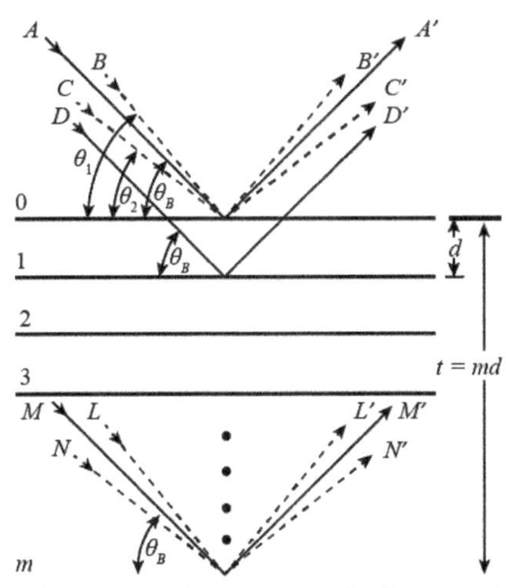

Schematic of crystal planes to calculate crystal size.

However, the intensity of the diffracted beam at angles close to $2\theta_B$, but not greater than $2\theta_1$ or less than $2\theta_2$ is not zero, but has an intermediate value between zero and one.

The width B as given in figure can roughly be measured from XRD at an intensity equal to half the maximum intensity, called as Full Width at Half Maximum (FWHM) and given as

$$B = \frac{1}{2}(2\theta_1 - 2\theta_2) = \theta_1 - \theta_2$$

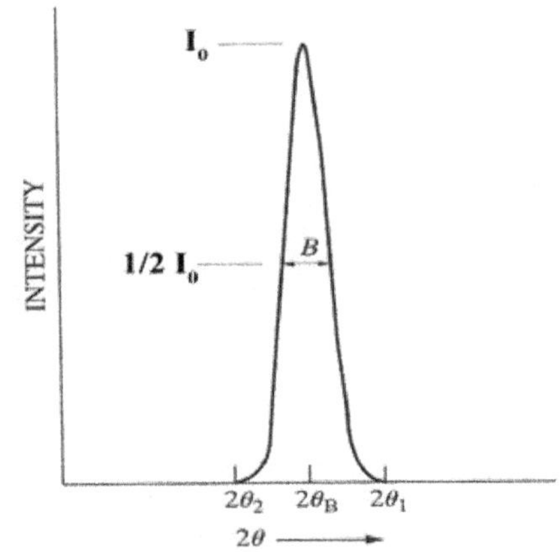

Schematic of a XRD peak with broadening.

The path difference equations for these two angles are

$$2\,t\,Sin\theta_1 = (m+1)\lambda$$
$$2\,t\,Sin\theta_2 = (m+1)\lambda$$

By subtracting, we find

$$t\,(sin\theta_1 - sin\theta_2) = \lambda$$

$$2\,t\cos\left(\frac{\theta_1+\theta_2}{2}\right)Sin\left(\frac{\theta_1+\theta_2}{2}\right) = \lambda$$

But θ_1 and θ_2 are nearly equal to θ_B, so that

$$\theta_1 + \theta_2 = 2\theta_B$$

and

$$\text{Sin}\left(\frac{\theta_1-\theta_2}{2}\right) = \left(\frac{\theta_1-\theta_2}{2}\right)$$

Therefore; $2t\cos\left(\frac{\theta_1+\theta_2}{2}\right)\text{Sin}\left(\frac{\theta_1-\theta_2}{2}\right) = \lambda$

$$2t\cos\theta_B\left(\frac{\theta_1-\theta_2}{2}\right) = \lambda$$

Or else

$$t = \frac{\lambda}{B\cos\theta_B}$$

with

$$B = \theta_1 - \theta_2 \Rightarrow \text{FWHM}$$

Mainly

$$t = \frac{k\lambda}{B\cos\theta_B}$$

where k is a constant representing shape factor. (k is 0.89 for spherical, 0.94 for cubic and 0.9 for unknown size particles).

This equation is known as Scherrer's formula for calculating the average crystallite size using XRD peaks. This equation not applicable for grains larger than 0.1 - 0.2 μm.

Calculation of d-spacing & Lattice Parameters

The various sets of planes in a lattice have different values of inter-planar spacing. The planes of larger spacing have low indices and pass through a high density of lattice points.

The inter-planar spacing d_{hkl} measured at right angles to the planes is a function of both the plane indices (*hkl*) and the lattice constants (a,b,c, α, β, γ), as shown in figure.

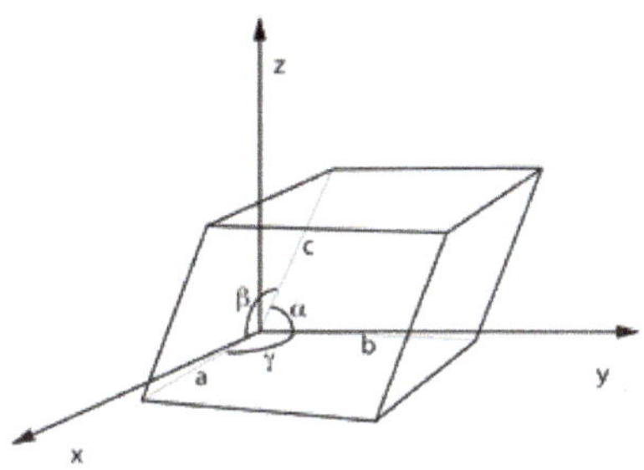

Schematic of a cube with different planes and angles.

The exact relation depends on the crystal system involved. The value of d can be calculated using Bragg's law

$$n\lambda = 2\,d\sin\theta$$

For the cubic systems as shown in figure, the relation between the lattice constant and d_{hkl} takes the form

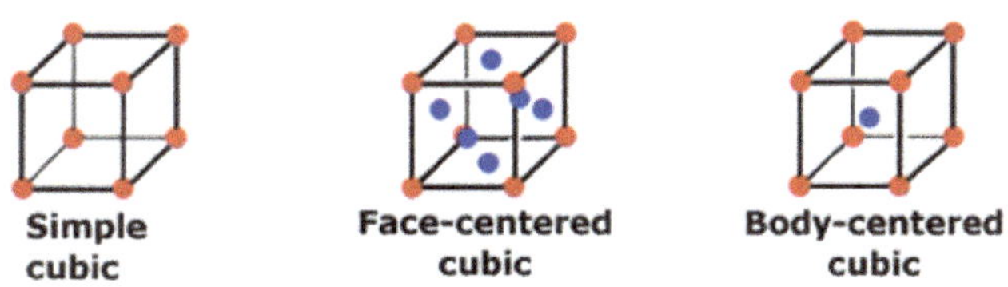

Schematic of different cubic systems.

$$d_{hkl} = \frac{\alpha}{\sqrt{h^2 + k^2 + l^2}}$$

In the tetragonal system, the spacing involves both a and c as they are not generally equal.

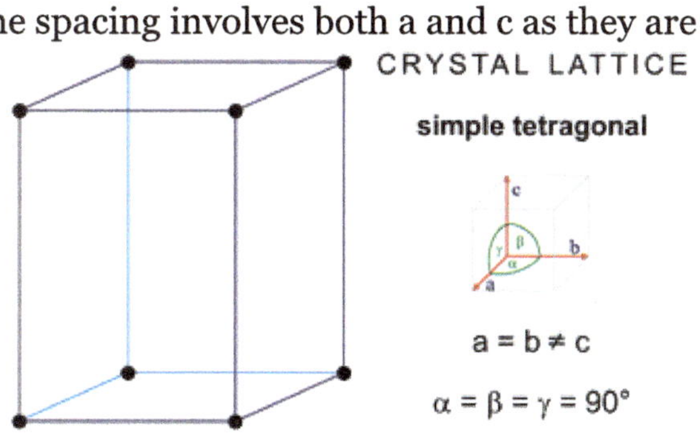

Schematic of tetragonal system.

Hence, the value of d_{hkl} takes different form,

$$d_{hkl} = \frac{a}{\sqrt{h^2 + k^2 + l^2 \frac{a^2}{c^2}}}$$

In the hexagonal system as displayed in figure, the spacing involves both a and c as they are not generally equal.

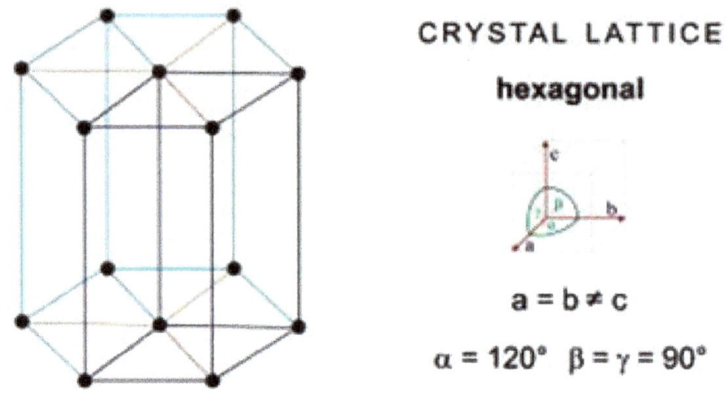

Schematic of hexagonal system.

Hence, the value of $d_{hk}l$ takes different form,

$$d_{hkl} = \frac{a}{\sqrt{\frac{4}{3}(h^2 + hk + k^2) l^2 \frac{a^2}{c^2}}}$$

Similarly, in the orthorhombic lattice system as shown in figure, the spacing involves all lengths (a, b and c) as they are not equal.

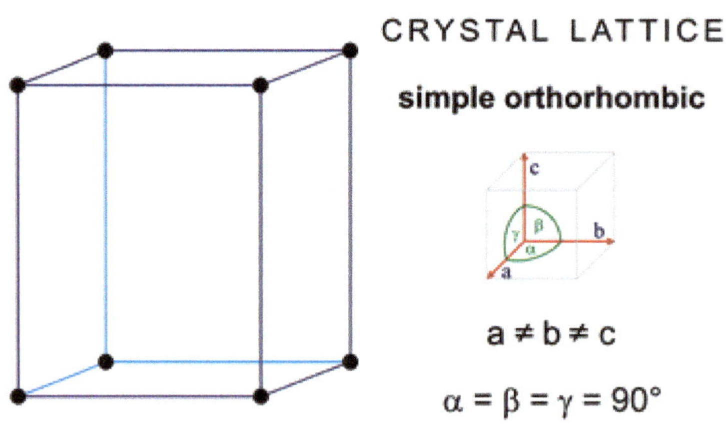

Schematic of orthorhombic system.

Hence, the value of d_{hkl} turns out to be,

$$\frac{1}{d_{hkl}^2} = \frac{h^2}{a^2} + \frac{k^2}{b^2} + \frac{l^2}{c^2}$$

The plane spacings of other lattice geometries are provided in the Appendix I of reference.

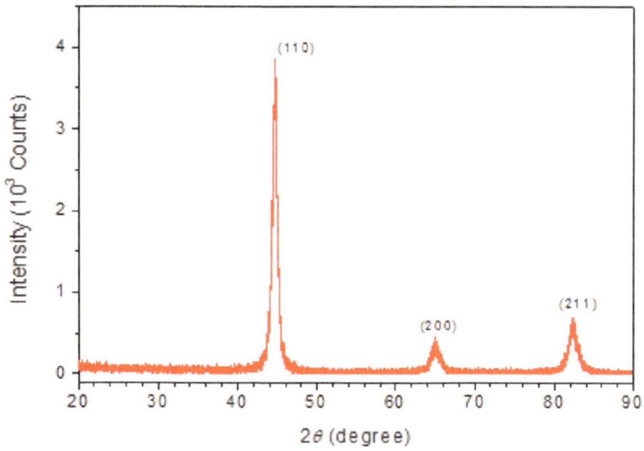

Calculate the average crystallite size, d-spacing, and lattice constant for the given body centered cubic structured polycrystalline material!

We need to find

(i) peak positions and

(ii) full width at half maxima for each peak.

Subsequently, we can calculate the average crystallite size, d - spacing and lattice constant by substituting these values in the below-mentioned equation.

$$2d\sin\theta = n\lambda \qquad t = \frac{k\lambda}{B\cos\theta}$$

$$a = d\sqrt{h^2 + k^2 + l^2}$$

Step 1: Procedure for finding peak positions and full width at Half-maxima *(FWHM)*.

Select the data for one of the peaks as shown in figure and execute the Gaussian single peak fit using any software (say, Origin).

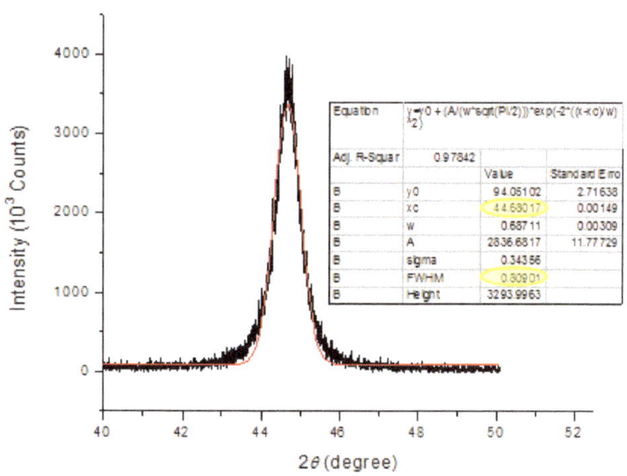

Fitting of XRD pattern using Gaussian fit and the resulting parameters.

The peak position and FWHM can be directly obtained from xc and FWHM

The obtained values are

(1) Peak 1 (110); 2θ = 44.68°; FWHM = 0.80901°.

(2) Peak 2 (200); 2θ = 65.045°; FWHM = 1.50079°.

(3) Peak 3 (211); 2θ = 82.39°; FWHM = 1.5103°.

Peaks	Interplanar spacing (d) $2d\sin\theta = n\lambda$	Lattice constant (a) $a = d\sqrt{h^2+k^2+l^2}$	Crystallite size $t = \frac{k\lambda}{B\cos\theta}$
(110)	2.0265 Å	2.8659 Å	10.64 nm
(200)	1.433 Å	2.8660 Å	6.28 nm
(211)	1.17 Å	2.8659 Å	6.99 nm

While the values of d decreases with increasing the diffraction angle, the lattice constant remains almost constant, irrespective of the peak we use for analysis.

Furthermore, the crystallite size varies significantly with the peaks. Considering the majority of the planes [as (110)] based on the intensity in the XRD pattern, the average crystallite size can be quantified as simply as 10.6 nm.

Effect of Stress

We have so far assumed that the peak broadening is mainly due to the reduction in the crystallite size.

Hence, the average crystallite size was calculated using Scherrer's formula.

It is also observed the peak broadening occurs due to the stress in the material, induced either during the sample fabrication process or during measurement conditions.

We shall now understand the effect of stress on the XRD peak and quantifying their contribution through proper (limited) analysis.

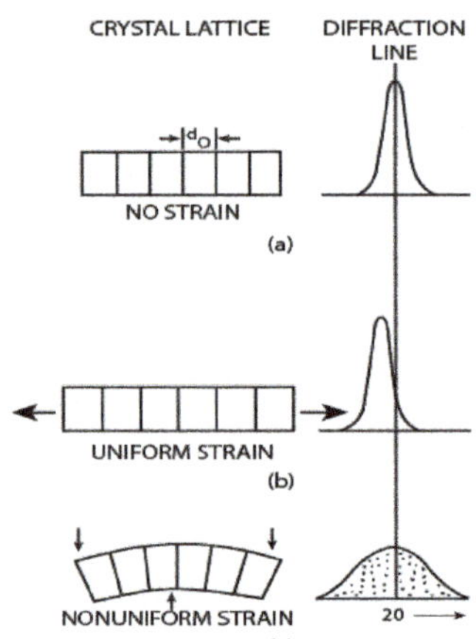

Schematic of effect of uniform and non-uniform strain in XRD peak.

Scherrer Equation

The Scherrer equation, in X-ray diffraction and crystallography, is a formula that relates the size of sub-micrometre particles, or crystallites, in a solid to the broadening of a peak in a diffraction pattern. It is named after Paul Scherrer. It is used in the determination of size of particles of crystals in the form of powder.

The Scherrer equation can be written as:

$$\tau = \frac{K\lambda}{\beta \cos\theta}$$

where:

- τ is the mean size of the ordered (crystalline) domains, which may be smaller or equal to the grain size;
- K is a dimensionless shape factor, with a value close to unity. The shape factor has a typical value of about 0.9, but varies with the actual shape of the crystallite;
- λ is the X-ray wavelength;
- β is the line broadening at half the maximum intensity (FWHM), after subtracting the instrumental line broadening, in radians. This quantity is also sometimes denoted as $\Delta(2\theta)$;
- θ is the Bragg angle (in degrees).

Applicability

The Scherrer equation is limited to nano-scale particles. It is not applicable to grains larger than about 0.1 to 0.2 μm, which precludes those observed in most metallographic and ceramographic microstructures.

It is important to realize that the Scherrer formula provides a lower bound on the particle size. The reason for this is that a variety of factors can contribute to the width of a diffraction peak besides instrumental effects and crystallite size; the most important of these are usually inhomogeneous strain and crystal lattice imperfections. The following sources of peak broadening are listed in reference: dislocations, stacking faults, twinning, microstresses, grain boundaries, sub-boundaries, coherency strain, chemical heterogeneities, and crystallite smallness. (Some of those and other imperfections may also result in peak shift, peak asymmetry, anisotropic peak broadening, or affect peak shape.)

If all of these other contributions to the peak width were zero, then the peak width would be determined solely by the crystallite size and the Scherrer formula would apply. If the other contributions to the width are non-zero, then the crystallite size can be larger than that predicted by the Scherrer formula, with the "extra" peak width coming from the other factors. The concept of crystallinity can be used to collectively describe the effect of crystal size and imperfections on peak broadening.

Derivation for a Simple Stack of Planes

To see where the Scherrer equation comes from, it is useful to consider the simplest possible example:a set of N planes spaced a apart. The derivation for this simple, effectively one-dimensional case, is straightforward. First we will derive the structure factor for this case, and then determine an expression for the peak widths.

Structure Factor for a Set of *N* Equally Spaced Planes

This system, effectively a one dimensional perfect crystal, has a structure factor or scattering function *S(q)*

$$S(q) = \frac{1}{N}\sum_{j,k=1}^{N} e^{-iq(x_j - x_k)}$$

where for *N* planes, $x_j = aj$, so

$$S(q) = \frac{1}{N}\sum_{k=1}^{N} e^{-iqak} \times \sum_{j=1}^{N} e^{iqaj}$$

the sums are simple geometric series, defining $y = \exp(iqa)$, $\sum_{j=1}^{N} y^j = (y - y^{N+1})/(1-y)$, with the other series analogous. Then

$$S(q) = \frac{1}{N}\frac{\left[e^{-iqa} - e^{-iqa(N+1)}\right]}{\left[1 - e^{-iqa}\right]} \times \frac{\left[e^{iqa} - e^{iqa(N+1)}\right]}{\left[1 - e^{iqa}\right]}$$

$$S(q) = \frac{1}{N}\frac{2 - e^{iqaN} - e^{-iqaN}}{2 - e^{iqa} - e^{-iqa}}$$

converting to trigonometric functions

$$S(q) = \frac{1}{N}\frac{1 - \cos[Nqa]}{1 - \cos[qa]}$$

and finally

$$S(q) = \frac{1}{N}\frac{\sin^2[Nqa/2]}{\sin^2[qa/2]}$$

which gives a set of peaks at $q_P = 0, 2\pi/a, 4\pi/a, \ldots$, all with heights $S(q_P) = N$.

Determination of the Profile Near the Peak, and Hence the Peak Width

From the definition of FWHM, for a peak at q_P and with a FWHM of Δq, $S(q_P \pm \Delta q) = S(q_P)/2 = N/2$, as the peak height is *N*. If we take the plus sign (peak is symmetric so either sign will do)

$$S(q_P + \Delta q) = \frac{1}{N}\frac{\sin^2[Na(q_P + \Delta q/2)/2]}{\sin^2[a(q_P + \Delta q/2)/2]} = N/2$$

and

$$\frac{\sin[Na(q_P + \Delta q/2)/2]}{\sin[a(q_P + \Delta q/2)/2]} = \frac{\sin[Na\Delta q/4]}{\sin[a\Delta q/4]} = \frac{N}{2^{1/2}}$$

as Δq is small for not-too-small *N*, then $\sin[\Delta qa/2] \simeq \Delta qa/2$, and we can write the equation is a single non-linear equation $\sin(x) - (x/2^{1/2}) = 0$, for $x = Na\Delta q/4$. The solution to this equation is $x = 1.39$. Therefore, the size of the set of planes is related to the FWHM in *q* by

$$\tau = Na = \frac{5.56}{\Delta q}$$

To convert to an expression for crystal size in terms of the peak width in the scattering angle 2θ used in X-ray powder diffraction, we note that the scattering vector $q = (4\pi/\lambda)\sin(\theta)$. Then the peak width in the variable $q = (4\pi/\lambda)\sin(\theta)$ is approximately $\beta \simeq 2\Delta q/[dq/d\theta] = 2\Delta q/[(4\pi/\lambda)\cos(\theta)]$, and so

$$\tau = Na = \frac{5.56\lambda}{2\pi\beta\cos(\theta)} = \frac{0.88\lambda}{\beta\cos(\theta)}$$

which is the Scherrer equation with $K = 0.88$.

This is all for 1D set of planes. In the experimentally relevant case of 3D, the form of $S(q)$ and hence the peaks, depends on the crystal lattice type, and the size and shape of the nanocrystallite. The maths is also more involved than in this simple illustrative example. However, for simple lattices and shapes, expressions have been obtained for the FWHM, for example by Patterson. Just as in 1D, the FWHM varies as one oversize. For example, for a spherical particle with a cubic lattice, the factor of 5.56 simply becomes 6.96, when the size is the diameter *D*, i.e., the diameter of a spherical nanocrystal is related to the peak FWHM by

$D = \frac{6.96}{\Delta q}$ or in θ: $D = \frac{1.11\lambda}{\beta\cos(\theta)}$

Peak Broadening Due to Disorder of the Second Kind

The finite size of a crystal is not the only possible reason for broadened peaks in X-ray diffraction. Fluctuations of atoms about the ideal lattice positions that preserve the long-range order of the lattice only give rise to the Debye-Waller factor, which reduces peak heights but does not broaden them. However, fluctuations that cause the correlations between nearby atoms to decrease as their separation increases, does broaden peaks. This can studied and quantified using the same simple one-dimensional stack of planes as above. The derivation follows that in chapter 9 of Guinier's textbook. This model has been pioneered by and applied to a number of materials bu Hosemann and collaborators over a number of years. They termed this disorder of the second kind, and referred to these imperfect crystalline ordering as paracrystalline ordering. Disorder of the first kind is the source of the Debye-Waller factor.

To derive the model we start with the definition of the structure factor

$$S(q) = \frac{1}{N}\sum_{j,k=1}^{N} e^{-iq(x_j - x_k)}$$

but now we want to consider, for simplicity an infinite crystal, i.e., $N \to \infty$, and we want to consider pairs of lattice sites. For large N, for each of these N planes, there are two neighbours m planes away, so the above double sum becomes a single sum over pairs of neighbours either side of an atom, at positions $-m$ and m lattice spacings away, times N. So, then

$$S(q) = 1 + \frac{2}{N}\sum_{m=1}^{N}\int_{-\infty}^{\infty} d(\Delta x) p_m(\Delta x)\cos(mq\Delta x)$$

where $p_m(\Delta x)$ is the probability density function for the separation Δx of a pair of planes, m lattice spacings apart. For the separation of neighbouring planes we assume for simplicity that the fluctuations around the mean neighbour spacing of a are Gaussian, i.e., that

$$p_1(\Delta x) = \frac{1}{\left(2\pi\sigma_2^2\right)^{1/2}}\exp\left[-(\Delta x - a)^2/(2\sigma_2^2)\right]$$

and we also assume that the fluctuations between a plane and its neighbour, and between this neighbour and the next plane, are independent. Then $p_2(\Delta x)$ is just the convolution of two $p_1(\Delta x)$ s, etc. As the convolution of two Gaussians is just another Gaussian, we have that

$$p_m(\Delta x) = \frac{1}{\left(2\pi m\sigma_2^2\right)^{1/2}} \exp\left[-(\Delta x - ma)^2 / (2m\sigma_2^2)\right]$$

The sum in $S(q)$ is then just a sum of Fourier Transforms of Gaussians, and so

$$S(q) = 1 + 2\sum_{m=1}^{\infty} r^m \cos(mqa)$$

for $r = \exp[-q^2\sigma_2^2/2]$. The sum is just the real part of the sum $\sum_{m=1}^{\infty}[\tau\exp(iqa)]^m$ and so the structure factor of the infinite but disordered crystal is

$$S(q) = \frac{1-r^2}{1+r^2-2r\cos(qa)}$$

This has peaks at maxima $q_p = 2n\pi/a$, , where $\cos(q_P a) = 1$. These peaks have heights

$$S(q_P) = \frac{1+r}{1-r} \approx \frac{4}{q_P^2\sigma_2^2} = \frac{1}{n^2\pi^2\sigma_2^2}$$

i.e., the height of successive peaks drop off as the order of the peak (and so q) squared. Unlike finite-size effects that broaden peaks but do not decrease their height, disorder lowers peak heights. Note that here we assuming that the disorder is relatively weak, so that we still have relatively well defined peaks. This is the limit $q\sigma_2 \ll 1$, where $r \simeq 1 - q^2\sigma_2^2/2$. In this limit, near a peak we can approximate $\cos(qa) \simeq 1 - (\Delta q)^2 a^2/2$, with $\Delta q = q - q_P$ and obtain

$$S(q) \approx \frac{S(q_P)}{1 + \frac{r}{(1-r)^2}\frac{\Delta q^2 a^2}{2}} \approx \frac{S(q_P)}{1 + \frac{\Delta q^2}{[q_P^2\sigma_2^2/a]^2/2}}$$

which is a Lorentzian or Cauchy function, of FWHM $q_P^2\sigma_2^2/a = 4\pi^2 n^2(\sigma_2/a)^2/a$, i.e., the FWHM increases as the square of the order of peak, and so as the square of the wavector q at the peak. Finally, the product of the peak height and the FWHM is constant and equals $4/a$, in the $q\sigma_2 \ll 1$ limit. For the first few peaks where n is not large, this is just the $\sigma_2/a \ll 1$ limit.

Thus finite-size and this type of disorder both cause peak broadening, but there are qualitative differences. Finite-size effects broadens all peaks equally, and does not affect peak heights, while this type of disorder both reduces peak heights and broadens peaks by an amount that increases as n^2. This, in principle, allows the two effects to be distinguished. Also, it means that the Scherrer equation is best applied to the first peak, as disorder of this type affects the first peak the least.

Coherence Length

Within this model the degree of correlation between a pair of planes decreases as the distance between these planes increases, i.e., a pair of planes 10 planes apart have positions that are more weakly correlated than a pair of planes that are nearest neighbours. The correlation is given by p_m, , for a pair of planes *m* planes apart. For sufficiently large *m* the pair of planes are essentially uncorrelated, in the sense that the uncertainty in their relative positions is so large that it is comparable to the lattice spacing, *a*. This defines a correlation length, λ, defined as the separation when the width of p_m, which is $m^{1/2}\sigma_2$ equals *a*. This gives

$$\lambda = \frac{a^3}{\sigma_2^2}$$

which is in effect an order-of-magnitude estimate for the size of domains of coherent crystalline lattices. Note that the FWHM of the first peak scales as σ_2^2 / a^3, so the coherence length is approximately 1/FWHM for the first peak.

Residual Stress and Strain

- Strain in the crystal lattice will produce a distortion of the diffraction line. A typical case of uniform and non-uniform strain in the crystal lattice is shown in figure.
- Macrostrain causes the lattice parameters to change resulting in a peak shift. Glycolation or heating of clay minerals are examples of induced macrostrains.
- Tensile and compressive forces can produce microstrains resulting in a broadening of the diffraction peaks or peak asymmetry in some cases.
- Dislocations, vacancies, shear planes, etc., can produce microstress, which results in a distribution of peaks around the unstressed peak location, appearing like a crude broadening of the peak.

Williamson-Hall Plot (WHP) Method

Considering both size and strain effect, Williamson and Hall developed a method to analyze the XRD peak as given below.

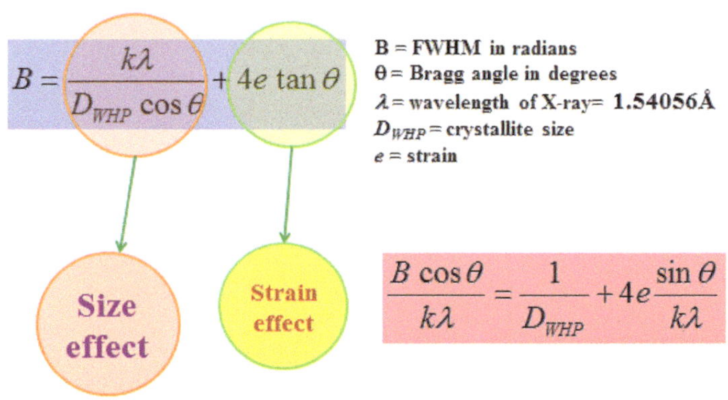

This is similar to y = mx + c. The plot of (Bcosθ/(kλ)) versus sinθ/(kλ) should be a straight line. Hence, the data can be fitted using a straight line, where the slope provides the strain in the sample and the intercept on the y-axis provides the inverse of average crystal size in the materials.

The applicability of the WHP method is described here. Figure displays the XRD patterns of the NiO sample prepared through sol-gel technique with different particle size and the analysis of XRD pattern using WHP method. These NiO particles exhibit face centered cubic structure.

1. All the peaks can be analyzed using Gaussian fit and each peak provides one FWHM and peak position.
2. Hence, five peaks in the XRD pattern provide five sets of ((Bcosθ /(kλ) , sinθ/(kλ)).
3. Therefore, the plot of (Bcosθ /(kλ)) versus sinθ/(kλ) reveals a straight line.

The determined value of strain in these particles is very small.

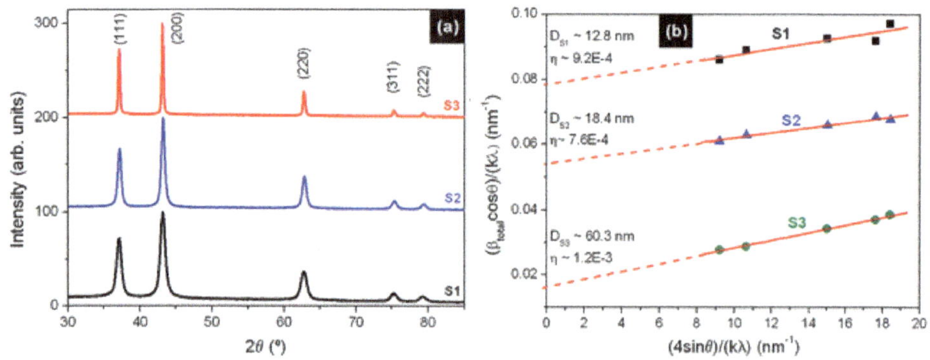

(a) Room temperature XRD patterns of NiO particles prepared using sol-gel method.
(b) Analysis of XRD pattern using WHP method.

However, the applicability of WHP method for the analysis of XRD pattern fails for the samples experience non-uniform strain. In such case, the plot of (Bcosθ /(kλ)) versus sinθ/(kλ) does not follow straight line. One such case is reported in figure for the Fe-Si powders prepared using mechanical alloying process.

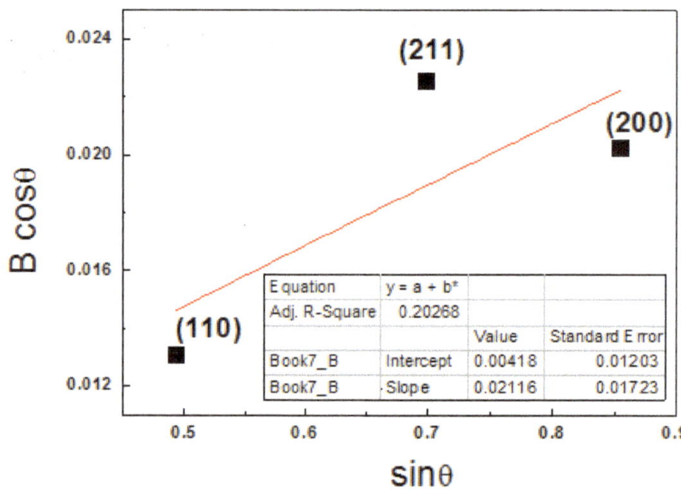

Non-applicability of WHP method for analyzing XRD pattern of Fe-Si alloy powders prepared by the mechanical alloying process.

The plot of B cosθ versus sinθ does not follow a straight line in some cases.

It is well-known that the alloy powders prepared using mechanical alloying process experience non-uniform strain.

Modified Williamson-Hall Plot (MWHP) Method

The deviations are larger when the material being studied is elastically anisotropic, where the residual strains affect some Bragg reflections more than others. Ungar and co-workers suggested methods to take this into consideration through the introduction of a dislocation contrast factor C_{hkl}.

$$\frac{B\cos\theta}{k\lambda} = \frac{1}{D_{WHP}} + 4e\frac{\sin\theta}{k\lambda}$$

$$\Delta K = \frac{0.9}{D_{WHP}} + 2eK$$

$$\Delta K = \frac{B\cos\theta}{\lambda};\ K = \frac{\sin\theta}{\lambda}$$

$$(\Delta K)^2 = \left(\frac{0.9}{D_{MWHP}}\right)^2 + 4e^2K^2$$

Figure displays the different plots of XRD peaks data using MWHP method without taking into dislocation factors. It is evident that the data points are not falling in a simple straight line.

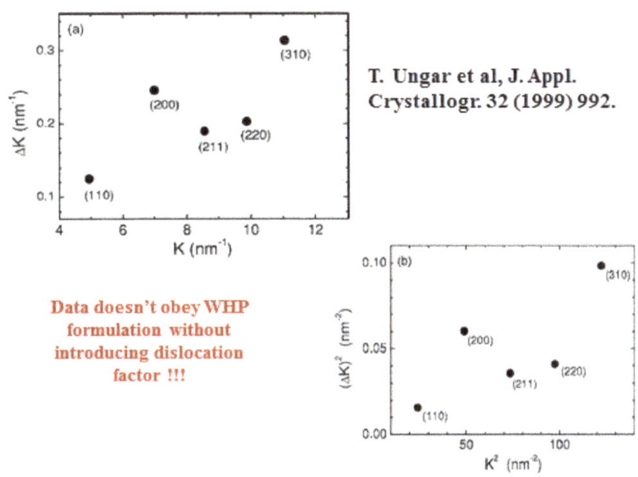

Different functional form of XRD peak data using MWHP method without the dislocation contrast factor.

After introducing the dislocation contrast factor into WHP formula, the equation turns out to be

$$\Delta K = \frac{0.9}{D_U} + 2eKC^{1/2}$$

$$(\Delta K)^2 = \left(\frac{0.9}{D_U}\right)^2 + 4e^2K^2C$$

$$(\Delta K)^2 = \left(\frac{0.9}{D_U}\right)^2 + \left(\frac{\pi b^2\rho}{2B}\right)^2 K^2C$$

$$C_{hkl} = C_{100}(1 - qH^2);\ H^2 = \frac{h^2k^2 + k^2l^2 + l^2 + h^2}{(h^2 + k^2 + l^2)^2}$$

where, D_U = crystallite size; ρ=dislocation density; C = C_{hkl} = dislocation contrast factor; b = Berger vector of dislocation; B = constant; θ_B = Bragg angle in degrees; $\Delta\theta_B$ = FWHM in radians; λ = wavelength of X-rays =1.54056Å); q = constant.

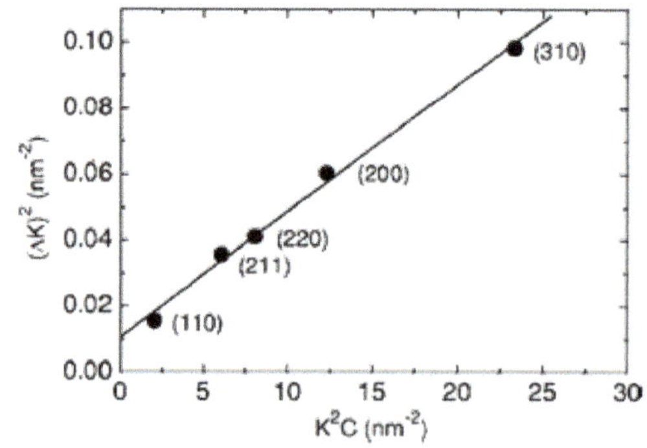

Analysis of XRD data using MWHP method after including dislocation contrast factor.

Figure shows the analysis of XRD data using MWHP method after including the dislocation contrast factor. It is clear from the figure that all the data points fall in a straight line without much deviation. Hence, the average crystal size and dislocation density causing strain in the sample can be determined from the slope and intercept of the straight line fit, respectively.

Preferred Orientation (Texture)

- The materials characteristics such as conductivity, thermal expansivity and hardness differ in each of the crystal directions.
- Most of the industrial materials possess polycrystalline nature and preferred orientation in the materials gives rise to any small change in the fabrication process.
- Therefore, the understanding of texture in the materials is very much important. This can be represented using pole figure, in which the existence of any number of crystallites in a specific crystallographic axis is plotted by a stereographic projection.
- A pole figure is closely examined whether or not the crystal grains in the plane are regularly aligned.
- Figure shows the diffractometer set up and sample movement in pole figure measurement. In the XRD set up, the sample and X-ray detector can be moved at θ and 2θ, respectively.
- After fixing θ or 2θ in a particular Bragg's condition for the sample as shown in figure, the pole figure measurement is undertaken by rotating the sample holder along the two-axes of Ψ and Φ. Ψ is varied at 0 - 90°, and Φ is adjusted from 0 - 360° for each value of Ψ as shown in figure.

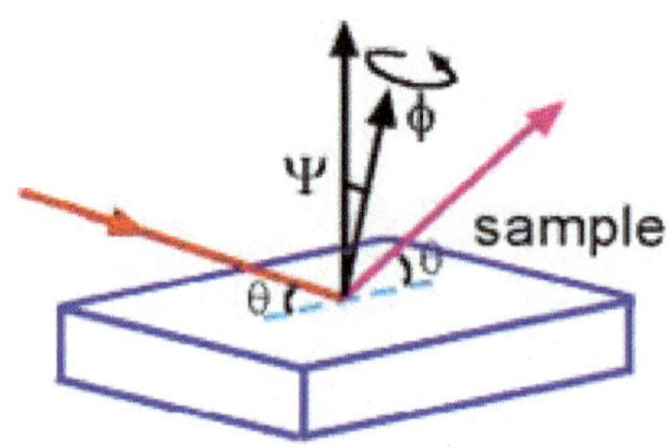

Pole Figure Measurement

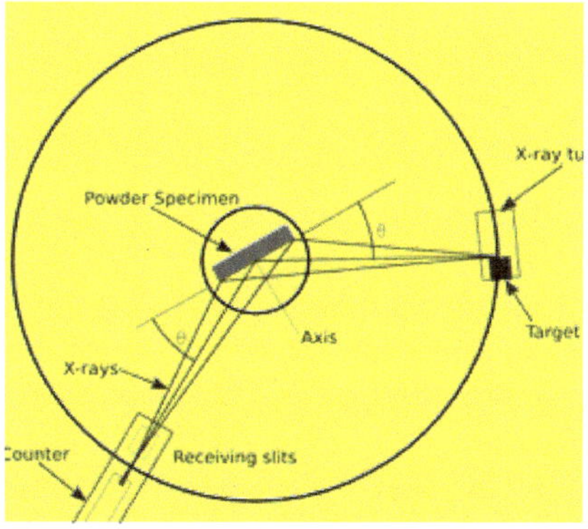

Schematic of XRD setup for texture analysis.

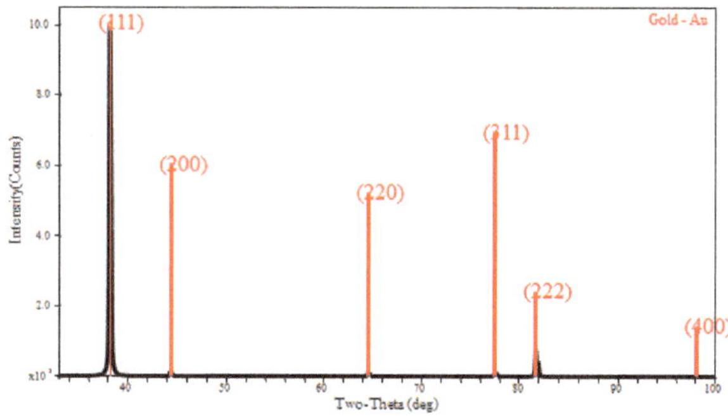

Typical XRD pattern of gold sample and selection of particular peak for pole figure analysis.

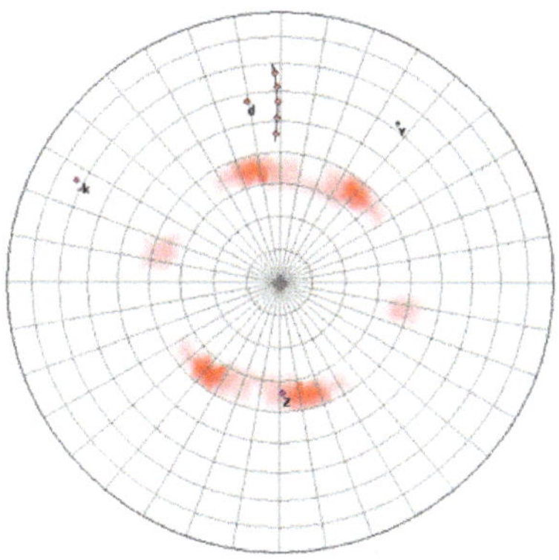

Pole figure obtained from XRD analysis.

- A pole figure is drawn as a contour map of X-ray intensity as a function of orientation of sample as shown in figure. X-ray intensity is proposed to the number of (hkl) planes in reflecting condition.
- If the crystal grains in the sample have random orientation, the pole figure exhibits a random intensity contour.

Neutron Diffraction

- Neutron diffraction or elastic neutron scattering is the application of neutron scattering to determine the atomic and/or magnetic structure of a material.

- A sample to be studied is placed in a beam of thermal or cold neutrons to obtain a diffraction pattern. The obtained patterns provide information of structure of the material.
- This technique is similar to X-ray diffraction but due to their different scattering properties, neutrons and X-rays provide complementary information.

History

- In 1932, neutron was first discovered by J. Chadwick
- In 1936, W. Elsasser proposed the idea of neutron scattering by crystalline materials.
- In 1936, F. Bloch predicted the feasibility of neutron scattering by magnetic moment in condensed materials.
- In 1940s and 1950s, high flux neutron reactor sources were built in U. S. and Canada (Chalk River's NRX reactor, ORNL's graphite reactor). In 1956, B. Brockhouse built the first triple-axis spectrometer in Chalk River Laboratory

E. Wollan and C. Shull did a lot of pioneering work in modern neutron diffraction between 1948 and 1955.

- 1994, the Nobel Prize

Bertram N. Brockhouse

Clifford G. Shull

Properties of X-ray and Neutron Scattering

The properties of X-rays and Neutrons are summarized in the table below:

Property	X-rays	Neutrons
Wavelength	Characteristic line spectra such as Cu K_α $\lambda = 1.54056$ Å	Continuous wavelength band, or single $\lambda = 1.1 \pm 0.05$ Å separated out from Maxwell spectrum by crystal monochromator or chopper
Energy for $\lambda = 1$ Å	10^{18} h	10^{13} h (same order as energy of elementary excitations)
Nature of scattering by atoms	Electronic Form factor dependence on $[\sin\theta]/\lambda$; Linear increase of scattering amplitude with atomic number, calculable from known electronic configurations	Nuclear, Isotropic, no angular dependent factor Irregular variation with atomic number. Dependent on nuclear structure and only determined empirically by experiment
Magnetic Scattering	Very weak additional scattering ($\approx 10^{-5}$)	Additional scattering by atoms with magnetic moments (same magnitude as nuclear scattering) Amplitude of scattering falls off with increasing $[\sin\theta]/\lambda$
Absorption coefficient	Very large, true absorption much larger than scattering $\mu_{abs} \approx 10^2 - 10^3$ increases with atomic number	Absorption usually very small (exceptions Gd, Cd, B ...) and less than scattering $\mu_{abs} \approx 10^{-1}$
Method of Detection	Solid State Detector, Image Plate	Proportional ^{3}He counter

Radiation: X-rays, Neutrons and Electrons

- Elastic scattering of radiation

  -- No energy is lost

- The wave length of the scattered wave remains unchanged
- Regular arrays of atoms interact elastically with radiation of sufficient short wavelength as shown in figure.

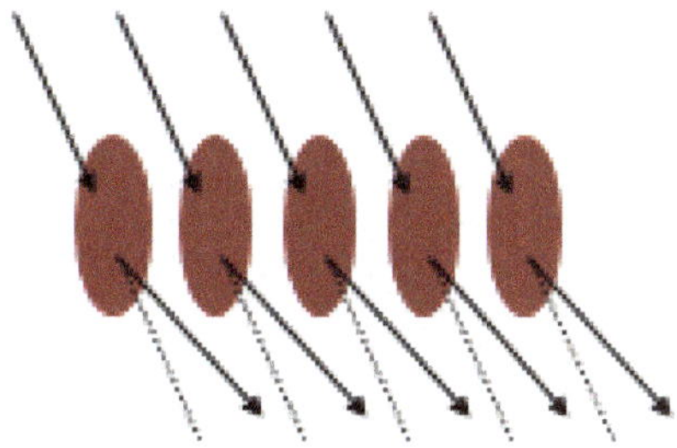

Schematic of interaction of radiation with the atom.

-- Cu-Kα X-ray radiation: $\lambda = 0.154$ nm

- Scattered by electrons
- ~from sub mm regions

-- Neutron radiation $\lambda \sim 0.1$nm

- Scattered by atomic nuclei
- Several cm thick samples

-- Electron radiation (200 kV): $\lambda = 0.00251$ nm

- Scattered by atomic nuclei and electrons
- Thickness less than 200 nm

Advantages of Neutron Diffraction

It is an important tool in the investigation of the magnetic ordering that occurs in some materials. Light atoms (say H) are better resolved in a neutron pattern. However, they do not contribute significantly to an X-Ray diffraction pattern.

Because they are neutral, neutrons do not interact with electrons in the crystal.So, unlike X-Rays, which are scattered entirely by electrons, neutrons are scattered entirely by nuclei.

Although uncharged, neutrons have an intrinsic magnetic moment, so they will interact strongly with atoms & ions in the crystal which also have a magnetic moment.

Neutrons are more useful than X-rays for determining the crystal structures of solids containing light elements.

Neutron sources in the world are limited so neutron diffraction is a very special tool & is very expensive.

Neutron was discovered in 1932 & their (de-Broglie) wave properties were proven in 1936.

Energy: $E = (p^2)/(2m_n)$

Momentum: $p = (h / \lambda)$

λ = Wavelength , $m_n = 1.67 \times 10^{-27}$ kg

Wavelengths needed for crystal diffraction are of the order of

$$\lambda \sim 1\text{Å}$$

So this gives neutron energies of

$$E \sim 0.08 \text{ eV}$$

This is of the same order of magnitude as the thermal energy at room temperature, $k_B T \sim 0.025$ eV. Hence, these are typically called as thermal neutrons.

Neutron Sources

There are different types of neutron sources available for experiments. Figure depicts two types of sources generated from continuous and pulsed sources.

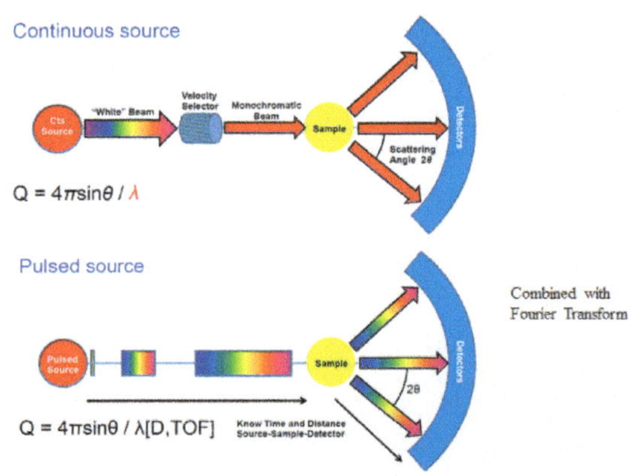

Neutron generation from continuous sources and pulsed sources.

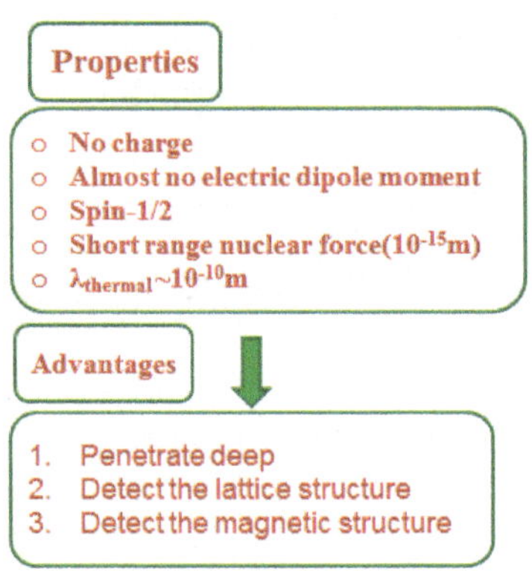

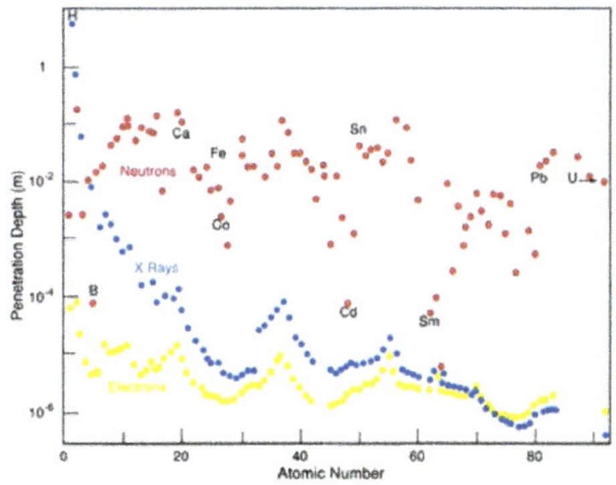

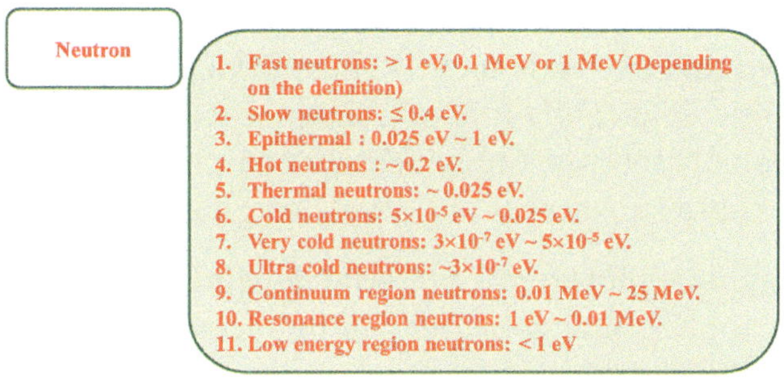

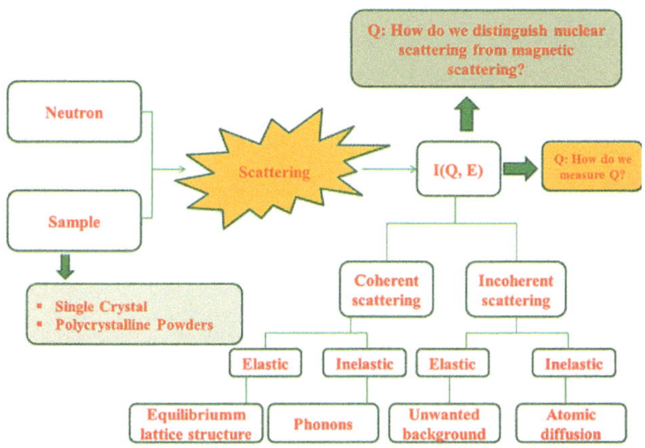

Magnetism

A magnetic quadrupole

Magnetism is a class of physical phenomena that are mediated by magnetic fields. Electric currents and the magnetic moments of elementary particles give rise to a magnetic field, which acts on other currents and magnetic moments. The most familiar effects occur in ferromagnetic materials, which are strongly attracted by magnetic fields and can be magnetized to become permanent magnets, producing magnetic fields themselves. Only a few substances are ferromagnetic; the most common ones are iron, nickel and cobalt and their alloys. The prefix *ferro-* refers to iron, because permanent magnetism was first observed in lodestone, a form of natural iron ore called magnetite, Fe_3O_4.

Although ferromagnetism is responsible for most of the effects of magnetism encountered in everyday life, all other materials are influenced to some extent by a magnetic field, by several other types of magnetism. Paramagnetic substances such as aluminum and oxygen are weakly attracted to an applied magnetic field; diamagnetic substances such as copper and carbon are weakly repelled; while antiferromagnetic materials such as chromium and spin glasses have a more complex relationship with a magnetic field. The force of a magnet on paramagnetic, diamagnetic, antiferromagnetic materials is usually too weak to be felt, and can be detected only by laboratory instruments, so in everyday life these substances are often described as non-magnetic.

The magnetic state (or magnetic phase) of a material depends on temperature and other variables such as pressure and the applied magnetic field. A material may exhibit more than one form of magnetism as these variables change.

History

Lodestone, a natural magnet, attracting iron nails. Ancient humans discovered the property of magnetism from lodestone.

An illustration from Gilbert's 1600 *De Magnete* showing one of the earliest methods of making a magnet. A blacksmith holds a piece of red-hot iron in a north-south direction and hammers it as it cools. The magnetic field of the Earth aligns the domains, leaving the iron a weak magnet.

Magnetism was first discovered in the ancient world, when people noticed that lodestones, naturally magnetized pieces of the mineral magnetite, could attract iron. The word *magnet* comes from the Greek term for lodestone, "magnítis líthos", which means a stone from the region of Magnesia. In ancient Greece, Aristotle attributed the first of what could be called a scientific discussion of magnetism to the philosopher Thales of Miletus, who lived from about 625 BC to about 545 BC. Around the same time, in ancient India, the Indian surgeon Sushruta was the first to make use of the magnet for surgical purposes.

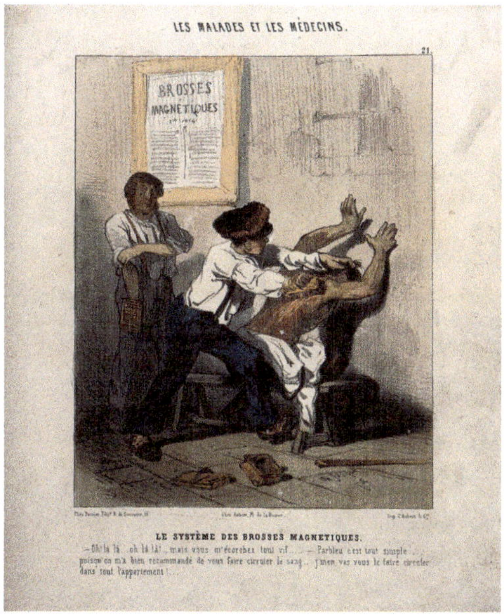

Drawing of a medical treatment using magnetic brushes. Charles Jacque 1843, France.

In ancient China, the earliest literary reference to magnetism lies in a 4th-century BC book named after its author, *The Master of Demon Valley*. The 2nd-century BC annals, *Lüshi Chunqiu*, also notes: "The lodestone makes iron approach, or it attracts it." The earliest mention of the attraction of a needle is in a 1st-century work *Lunheng* (*Balanced Inquiries*): "A lodestone attracts a needle." The 11th-century Chinese scientist Shen Kuo was the first person to write – in the *Dream Pool Essays* – of the magnetic needle compass and that it improved the accuracy of navigation by employing the astronomical concept of true north. By the 12th century the Chinese were known to use the lodestone compass for navigation. They sculpted a directional spoon from lodestone in such a way that the handle of the spoon always pointed south.

Alexander Neckam, by 1187, was the first in Europe to describe the compass and its use for navigation. In 1269, Peter Peregrinus de Maricourt wrote the *Epistola de magnete*, the first extant treatise describing the properties of magnets. In 1282, the properties of magnets and the dry compass were discussed by Al-Ashraf, a Yemeni physicist, astronomer, and geographer.

In 1600, William Gilbert published his *De Magnete, Magneticisque Corporibus, et de Magno Magnete Tellure* (*On the Magnet and Magnetic Bodies, and on the Great Magnet the Earth*). In this work he describes many of his experiments with his model earth called the terrella. From his experiments, he concluded that the Earth was itself magnetic and that this was the reason compasses pointed north (previously, some believed that it was the pole star (Polaris) or a large magnetic island on the north pole that attracted the compass).

An understanding of the relationship between electricity and magnetism began in 1819 with work by Hans Christian Ørsted, a professor at the University of Copenhagen, who discovered by the accidental twitching of a compass needle near a wire that an electric current could create a magnetic field. This landmark experiment is known as Ørsted's Experiment. Several other experiments followed, with André-Marie Ampère, who in 1820 discovered that the magnetic field circulating in a closed-path was related to the current flowing through the perimeter of the path; Carl Friedrich Gauss; Jean-Baptiste Biot and Félix Savart, both of whom in 1820 came up with the Biot–Savart law giving an equation for the magnetic field from a current-carrying wire; Michael Faraday, who in 1831 found that a time-varying magnetic flux through a loop of wire induced a voltage, and others finding further links between magnetism and electricity. James Clerk Maxwell synthesized and expanded these insights into Maxwell's equations, unifying electricity, magnetism, and optics into the field of electromagnetism. In 1905, Einstein used these laws in motivating his theory of special relativity, requiring that the laws held true in all inertial reference frames.

Electromagnetism has continued to develop into the 21st century, being incorporated into the more fundamental theories of gauge theory, quantum electrodynamics, electroweak theory, and finally the standard model.

Sources of Magnetism

Magnetism, at its root, arises from two sources:

1. Electric current.
2. Spin magnetic moments of elementary particles. The magnetic moments of the nuclei of atoms are typically thousands of times smaller than the electrons' magnetic moments, so they are negligible in the context of the magnetization of materials. Nuclear magnetic moments are nevertheless very important in other contexts, particularly in nuclear magnetic resonance (NMR) and magnetic resonance imaging (MRI).

Ordinarily, the enormous number of electrons in a material are arranged such that their magnetic moments (both orbital and intrinsic) cancel out. This is due, to some extent, to electrons combining into pairs with opposite intrinsic magnetic moments as a result of the Pauli exclusion principle, or combining into filled subshells with zero net orbital motion. In both cases, the electron arrangement is so as to exactly cancel the magnetic moments from each electron. Moreover, even when the electron configuration *is* such that there are unpaired electrons and/ or non-filled subshells, it is often the case that the various electrons in the solid will contribute magnetic moments that point in different, random directions, so that the material will not be magnetic.

Sometimes, either spontaneously, or owing to an applied external magnetic field—each of the electron magnetic moments will be, on average, lined up. A suitable material can then produce a strong net magnetic field.

The magnetic behavior of a material depends on its structure, particularly its electron configuration, for the reasons mentioned above, and also on the temperature. At high temperatures, random thermal motion makes it more difficult for the electrons to maintain alignment.

Materials

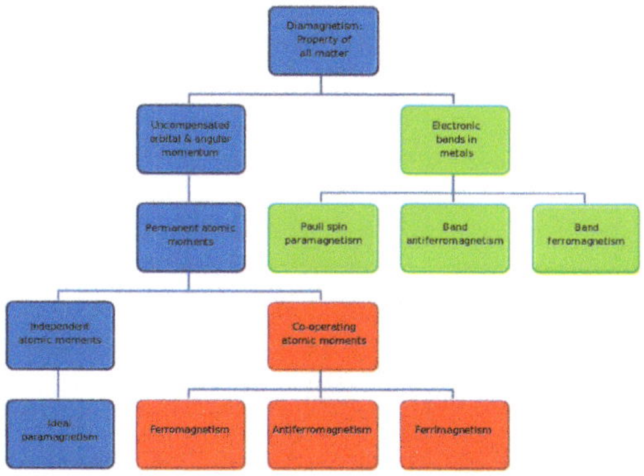

Hierarchy of types of magnetism.

Diamagnetism

Diamagnetism appears in all materials, and is the tendency of a material to oppose an applied magnetic field, and therefore, to be repelled by a magnetic field. However, in a material with paramagnetic properties (that is, with a tendency to enhance an external magnetic field), the paramagnetic behavior dominates. Thus, despite its universal occurrence, diamagnetic behavior is observed only in a purely diamagnetic material. In a diamagnetic material, there are no unpaired electrons, so the intrinsic electron magnetic moments cannot produce any bulk effect. In these cases, the magnetization arises from the electrons' orbital motions, which can be understood classically as follows:

When a material is put in a magnetic field, the electrons circling the nucleus will experience, in addition to their Coulomb attraction to the nucleus, a Lorentz force from the magnetic field. Depending on which direction the electron is orbiting, this force may increase the centripetal force on the electrons, pulling them in towards the nucleus, or it may decrease the force, pulling them away from the nucleus. This effect systematically increases the orbital magnetic moments that were aligned opposite the field, and decreases the ones aligned parallel to the field (in accordance with Lenz's law). This results in a small bulk magnetic moment, with an opposite direction to the applied field.

Note that this description is meant only as a heuristic; a proper understanding requires a quantum-mechanical description.

Note that all materials undergo this orbital response. However, in paramagnetic and ferromagnetic substances, the diamagnetic effect is overwhelmed by the much stronger effects caused by the unpaired electrons.

Paramagnetism

In a paramagnetic material there are *unpaired electrons*, i.e. atomic or molecular orbitals with exactly one electron in them. While paired electrons are required by the Pauli exclusion principle to have their intrinsic ('spin') magnetic moments pointing in opposite directions, causing their magnetic fields to cancel out, an unpaired electron is free to align its magnetic moment in any direction. When an external magnetic field is applied, these magnetic moments will tend to align themselves in the same direction as the applied field, thus reinforcing it.

Ferromagnetism

Tip of permanent magnet with coins demonstrating ferromagnetism

A ferromagnet, like a paramagnetic substance, has unpaired electrons. However, in *addition* to the electrons' intrinsic magnetic moment's tendency to be parallel to an *applied field*, there is also in these materials a tendency for these magnetic moments to orient parallel to *each other* to maintain a lowered-energy state. Thus, even in the absence of an applied field, the magnetic moments of the electrons in the material spontaneously line up parallel to one another.

Every ferromagnetic substance has its own individual temperature, called the Curie temperature, or Curie point, above which it loses its ferromagnetic properties. This is because the thermal tendency to disorder overwhelms the energy-lowering due to ferromagnetic order.

Ferromagnetism only occurs in a few substances; the common ones are iron, nickel, cobalt, their alloys, and some alloys of rare earth metals.

Magnetic Domains

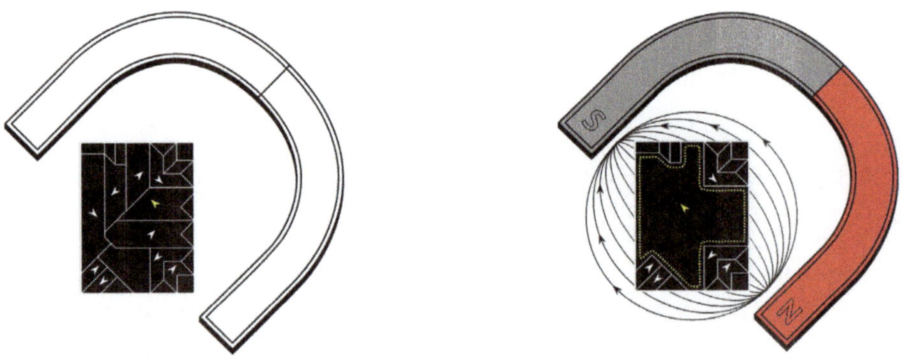

Left: Magnetic domains boundaries (white lines) in ferromagnetic material (black rectangle).
Right: Effect of a magnet on the domains.

The magnetic moments of atoms in a ferromagnetic material cause them to behave something like tiny permanent magnets. They stick together and align themselves into small regions of more or less uniform alignment called magnetic domains or Weiss domains. Magnetic domains can be observed with a magnetic force microscope to reveal magnetic domain boundaries that resemble white lines in the sketch. There are many scientific experiments that can physically show magnetic fields.

When a domain contains too many molecules, it becomes unstable and divides into two domains aligned in opposite directions so that they stick together more stably.

When exposed to a magnetic field, the domain boundaries move so that the domains aligned with the magnetic field grow and dominate the structure (dotted yellow area). When the magnetizing field is removed, the domains may not return to an unmagnetized state. This results in the ferromagnetic material's being magnetized, forming a permanent magnet.

When magnetized strongly enough that the prevailing domain overruns all others to result in only one single domain, the material is magnetically saturated. When a magnetized ferromagnetic material is heated to the Curie point temperature, the molecules are agitated to the point that the magnetic domains lose the organization and the magnetic properties they cause cease. When the material is cooled, this domain alignment structure spontaneously returns, in a manner roughly analogous to how a liquid can freeze into a crystalline solid.

Antiferromagnetism

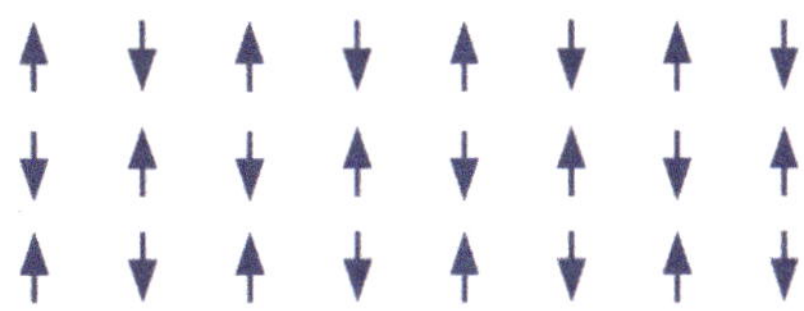

Antiferromagnetic ordering

In an antiferromagnet, unlike a ferromagnet, there is a tendency for the intrinsic magnetic moments of neighboring valence electrons to point in *opposite* directions. When all atoms are arranged in a substance so that each neighbor is 'anti-aligned', the substance is antiferromagnetic. Antiferromagnets have a zero net magnetic moment, meaning no field is produced by them. Antiferromagnets are less common compared to the other types of behaviors, and are mostly observed at low temperatures. In varying temperatures, antiferromagnets can be seen to exhibit diamagnetic and ferromagnetic properties.

In some materials, neighboring electrons want to point in opposite directions, but there is no geometrical arrangement in which *each* pair of neighbors is anti-aligned. This is called a spin glass, and is an example of geometrical frustration.

Ferrimagnetism

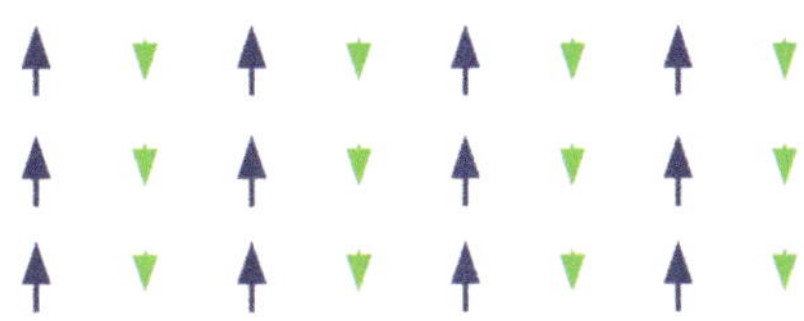

Ferrimagnetic ordering

Like ferromagnetism, ferrimagnets retain their magnetization in the absence of a field. However, like antiferromagnets, neighboring pairs of electron spins tend to point in opposite directions. These two properties are not contradictory, because in the optimal geometrical arrangement, there is more magnetic moment from the sublattice of electrons that point in one direction, than from the sublattice that points in the opposite direction.

Most ferrites are ferrimagnetic. The first discovered magnetic substance, magnetite, is a ferrite and was originally believed to be a ferromagnet; Louis Néel disproved this, however, after discovering ferrimagnetism.

Superparamagnetism

When a ferromagnet or ferrimagnet is sufficiently small, it acts like a single magnetic spin that is subject to Brownian motion. Its response to a magnetic field is qualitatively similar to the response of a paramagnet, but much larger.

Other Types of Magnetism

- Metamagnetism
- Molecule-based magnet
- Spin glass

Electromagnet

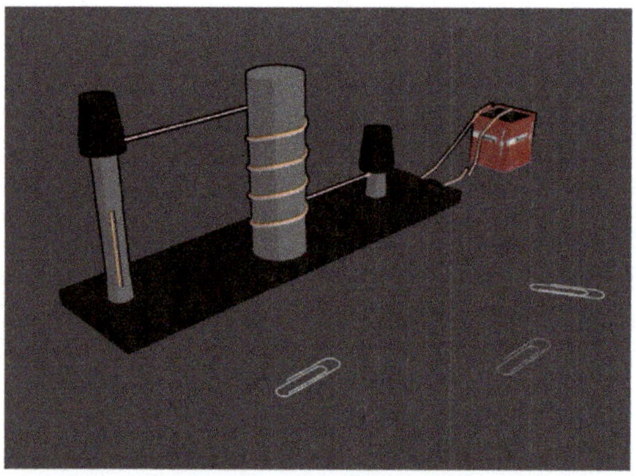

An electromagnet attracts paper clips when current is applied creating a magnetic field. The electromagnet loses them when current and magnetic field are removed.

An electromagnet is a type of magnet in which the magnetic field is produced by an electric current. The magnetic field disappears when the current is turned off. Electromagnets usually consist of a large number of closely spaced turns of wire that create the magnetic field. The wire turns are often wound around a magnetic core made from a ferromagnetic or ferrimagnetic material such as iron; the magnetic core concentrates the magnetic flux and makes a more powerful magnet.

The main advantage of an electromagnet over a permanent magnet is that the magnetic field can be quickly changed by controlling the amount of electric current in the winding. However, unlike a permanent magnet that needs no power, an electromagnet requires a continuous supply of current to maintain the magnetic field.

Electromagnets are widely used as components of other electrical devices, such as motors, generators, relays, loudspeakers, hard disks, MRI machines, scientific instruments, and magnetic separation equipment. Electromagnets are also employed in industry for picking up and moving heavy iron objects such as scrap iron and steel. Electromagnetism was discovered in 1820.

Magnetism, Electricity, and Special Relativity

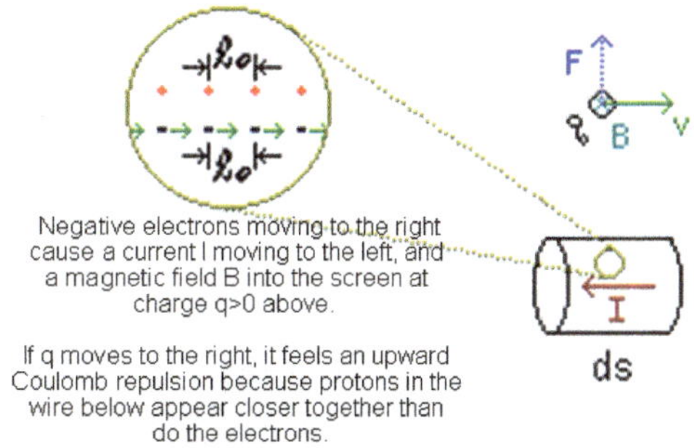

Magnetism from length-contraction.

As a consequence of Einstein's theory of special relativity, electricity and magnetism are fundamentally interlinked. Both magnetism lacking electricity, and electricity without magnetism, are inconsistent with special relativity, due to such effects as length contraction, time dilation, and the fact that the magnetic force is velocity-dependent. However, when both electricity and magnetism are taken into account, the resulting theory (electromagnetism) is fully consistent with special relativity. In particular, a phenomenon that appears purely electric or purely magnetic to one observer may be a mix of both to another, or more generally the relative contributions of electricity and magnetism are dependent on the frame of reference. Thus, special relativity "mixes" electricity and magnetism into a single, inseparable phenomenon called electromagnetism, analogous to how relativity "mixes" space and time into spacetime.

All observations on electromagnetism apply to what might be considered to be primarily magnetism, e.g. perturbations in the magnetic field are necessarily accompanied by a nonzero electric field, and propagate at the speed of light.

Magnetic Fields in a Material

In a vacuum,

$$\mathbf{B} = \mu_0 \mathbf{H},$$

where μ_0 is the vacuum permeability.

In a material,

$$\mathbf{B} = \mu_0 (\mathbf{H} + \mathbf{M}).$$

The quantity $\mu_0 M$ is called *magnetic polarization.*

If the field H is small, the response of the magnetization M in a diamagnet or paramagnet is approximately linear:

$$\mathbf{M} = \chi \mathbf{H},$$

the constant of proportionality being called the magnetic susceptibility. If so,

$$\mu_0 (\mathbf{H} + \mathbf{M}) = \mu_0 (1 + \chi) \mathbf{H} = \mu_r \mu_0 \mathbf{H} = \mu \mathbf{H}.$$

In a hard magnet such as a ferromagnet, M is not proportional to the field and is generally nonzero even when H is zero.

Magnetic Force

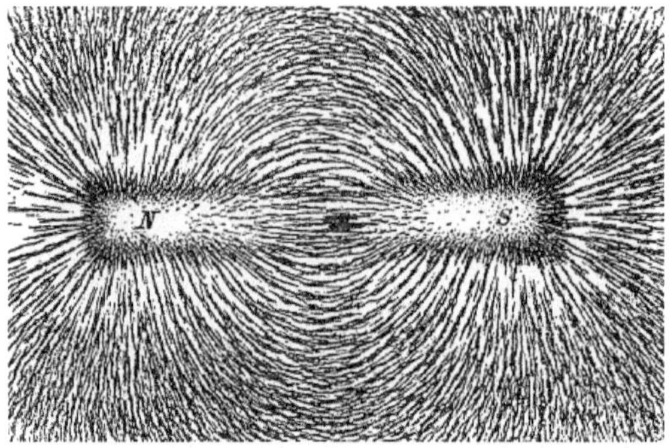

Magnetic lines of force of a bar magnet shown by iron filings on paper

The phenomenon of magnetism is "mediated" by the magnetic field. An electric current or magnetic dipole creates a magnetic field, and that field, in turn, imparts magnetic forces on other particles that are in the fields.

Maxwell's equations, which simplify to the Biot–Savart law in the case of steady currents, describe the origin and behavior of the fields that govern these forces. Therefore, magnetism is seen whenever electrically charged particles are in motion—for example, from movement of electrons in an electric current, or in certain cases from the orbital motion of electrons around an atom's nucleus. They also arise from "intrinsic" magnetic dipoles arising from quantum-mechanical spin.

The same situations that create magnetic fields—charge moving in a current or in an atom, and intrinsic magnetic dipoles—are also the situations in which a magnetic field has an effect, creating a force. Following is the formula for moving charge; for the forces on an intrinsic dipole.

When a charged particle moves through a magnetic field B, it feels a Lorentz force F given by the cross product:

$$\mathbf{F} = q(\mathbf{v} \times \mathbf{B})$$

where

q is the electric charge of the particle, and

v is the velocity vector of the particle

Because this is a cross product, the force is perpendicular to both the motion of the particle and the magnetic field. It follows that the magnetic force does no work on the particle; it may change the direction of the particle's movement, but it cannot cause it to speed up or slow down. The magnitude of the force is

$$F = qvB\sin\theta$$

where θ is the angle between v and B.

One tool for determining the direction of the velocity vector of a moving charge, the magnetic field, and the force exerted is labeling the index finger "V", the middle finger "B", and the thumb "F" with your right hand. When making a gun-like configuration, with the middle finger crossing under the index finger, the fingers represent the velocity vector, magnetic field vector, and force vector, respectively.

Magnetic Dipoles

A very common source of magnetic field found in nature is a dipole, with a "South pole" and a "North pole", terms dating back to the use of magnets as compasses, interacting with the Earth's magnetic field to indicate North and South on the globe. Since opposite ends of magnets are attracted, the north pole of a magnet is attracted to the south pole of another magnet. The Earth's North Magnetic Pole (currently in the Arctic Ocean, north of Canada) is physically a south pole, as it attracts the north pole of a compass. A magnetic field contains energy, and physical systems move toward configurations with lower energy. When diamagnetic material is placed in a magnetic field, a *magnetic dipole* tends to align itself in opposed polarity to that field, thereby lowering the net field strength. When ferromagnetic material is placed within a magnetic field, the magnetic dipoles align to the applied field, thus expanding the domain walls of the magnetic domains.

Magnetic Monopoles

Since a bar magnet gets its ferromagnetism from electrons distributed evenly throughout the bar, when a bar magnet is cut in half, each of the resulting pieces is a smaller bar magnet. Even though a magnet is said to have a north pole and a south pole, these two poles cannot be separated from each other. A monopole—if such a thing exists—would be a new and fundamentally different kind of magnetic object. It would act as an isolated north pole, not attached to a south pole, or vice versa. Monopoles would carry "magnetic charge" analogous to electric charge. Despite systematic searches since 1931, as of 2010, they have never been observed, and could very well not exist.

Nevertheless, some theoretical physics models predict the existence of these magnetic monopoles. Paul Dirac observed in 1931 that, because electricity and magnetism show a certain symmetry, just as quantum theory predicts that individual positive or negative electric charges can be observed without the opposing charge, isolated South or North magnetic poles should be observable. Using quantum theory Dirac showed that if magnetic monopoles exist, then one could explain the quantization of electric charge—that is, why the observed elementary particles carry charges that are multiples of the charge of the electron.

Certain grand unified theories predict the existence of monopoles which, unlike elementary particles, are solitons (localized energy packets). The initial results of using these models to estimate the number of monopoles created in the big bang contradicted cosmological observations—the monopoles would have been so plentiful and massive that they would have long since halted the expansion of the universe. However, the idea of inflation (for which this problem served as a partial motivation) was successful in solving this problem, creating models in which monopoles existed but were rare enough to be consistent with current observations.

Quantum-mechanical Origin of Magnetism

In principle all kinds of magnetism originate from specific quantum-mechanical phenomena (e.g. Mathematical formulation of quantum mechanics, in particular the chapters on spin and on the Pauli principle). A successful model was developed already in 1927, by Walter Heitler and Fritz

London, who derived quantum-mechanically, how hydrogen molecules are formed from hydrogen atoms, i.e. from the atomic hydrogen orbitals u_A and u_B centered at the nuclei *A* and *B*. That this leads to magnetism is not at all obvious, but will be explained in the following.

According to the Heitler-London theory, so-called two-body molecular σ-orbitals are formed, namely the resulting orbital is:

$$\psi(\mathbf{r}_1,\mathbf{r}_2)=\frac{1}{\sqrt{2}}(u_A(\mathbf{r}_1)u_B(\mathbf{r}_2)+u_B(\mathbf{r}_1)u_A(\mathbf{r}_2))$$

Here the last product means that a first electron, r_1, is in an atomic hydrogen-orbital centered at the second nucleus, whereas the second electron runs around the first nucleus. This "exchange" phenomenon is an expression for the quantum-mechanical property that particles with identical properties cannot be distinguished. It is specific not only for the formation of chemical bonds, but also for magnetism, i.e. in this connection the term exchange interaction arises, a term which is essential for the origin of magnetism, and which is stronger, roughly by factors 100 and even by 1000, than the energies arising from the electrodynamic dipole-dipole interaction.

As for the *spin function* $\chi(s_1,s_2)$, which is responsible for the magnetism, we have the already mentioned Pauli's principle, namely that a symmetric orbital (i.e. with the + sign as above) must be multiplied with an antisymmetric spin function (i.e. with a – sign), and *vice versa*. Thus:

$$\chi(s_1,s_2)=\frac{1}{\sqrt{2}}(\alpha(s_1)\beta(s_2)-\beta(s_1)\alpha(s_2)),$$

I.e., not only u_A and u_B must be substituted by α and β, respectively (the first entity means "spin up", the second one "spin down"), but also the sign + by the – sign, and finally r_i by the discrete values s_i (= ±½); thereby we have $\alpha(+1/2)=\beta(-1/2)=1$ and $\alpha(-1/2)=\beta(+1/2)=0$. The "singlet state", i.e. the – sign, means: the spins are *antiparallel*, i.e. for the solid we have antiferromagnetism, and for two-atomic molecules one has diamagnetism. The tendency to form a (homoeopolar) chemical bond (this means: the formation of a *symmetric* molecular orbital, i.e. with the + sign) results through the Pauli principle automatically in an *antisymmetric* spin state (i.e. with the – sign). In contrast, the Coulomb repulsion of the electrons, i.e. the tendency that they try to avoid each other by this repulsion, would lead to an *antisymmetric* orbital function (i.e. with the – sign) of these two particles, and complementary to a *symmetric* spin function (i.e. with the + sign, one of the so-called "triplet functions"). Thus, now the spins would be *parallel* (ferromagnetism in a solid, paramagnetism in two-atomic gases).

The last-mentioned tendency dominates in the metals iron, cobalt and nickel, and in some rare earths, which are *ferromagnetic*. Most of the other metals, where the first-mentioned tendency dominates, are *nonmagnetic* (e.g. sodium, aluminium, and magnesium) or *antiferromagnetic* (e.g. manganese). Diatomic gases are also almost exclusively diamagnetic, and not paramagnetic. However, the oxygen molecule, because of the involvement of π-orbitals, is an exception important for the life-sciences.

The Heitler-London considerations can be generalized to the Heisenberg model of magnetism (Heisenberg 1928).

The explanation of the phenomena is thus essentially based on all subtleties of quantum mechanics, whereas the electrodynamics covers mainly the phenomenology.

(Magnetic) Neutron Scattering

Many features of thermal neutrons make them ideal for crystal structure determination:

1. Weakly interacting
2. Penetrating
3. Non-destructive
4. Wavelength matches lattice spacing
5. Light/heavy element sensitivity
    - Neutron also has a magnetic moment, i.e., it can scatter from other magnetic moments in the sample.œ
    - The magnetic scattering length is comparable to the nuclear scattering length, i.e., magnetic and nuclear signals have comparable strengths.

Difference between Magnetic and Nuclear scattering:

- They normally occur at different wave vectors.
- While magnetic scattering is temperature dependent, the nuclear scattering does not depend on temperature.
- Using polarized neutrons, spin flipping can be obtained for magnetic scattering.

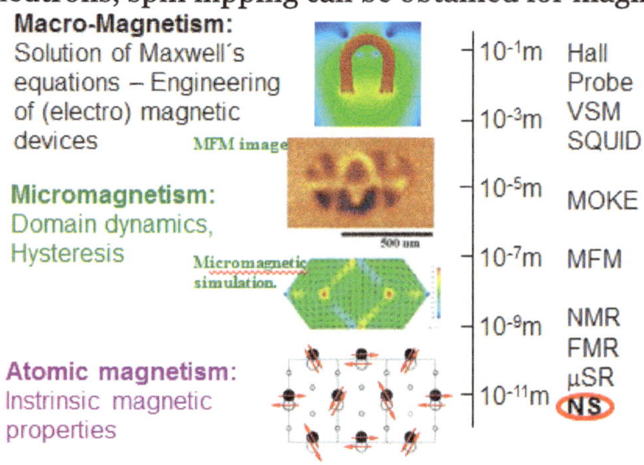

Magnetic Structure Determination

- Everything that you have heard about crystal structure determination applies equally to the magnetic structure determination.
- Neutron scattering pattern observed for a magnetic material is just the Fourier transform of the nuclear and magnetic structure.
- Therefore, the information we learn about the crystal structure can also be learned about the magnetic structure. Almost the same basic analysis and principles are followed.

- Neutron diffraction is the best way to study magnetic ordering.
- Should always be combined with complementary methods to obtain a more complete picture

Examples Include

- Bulk magnetic methods (magnetization, susceptibility)
- Mössbauer spectroscopy
- Nuclear Magnetic Resonance
- Muon spin spectroscopy
- Time-differential perturbed angular correlation

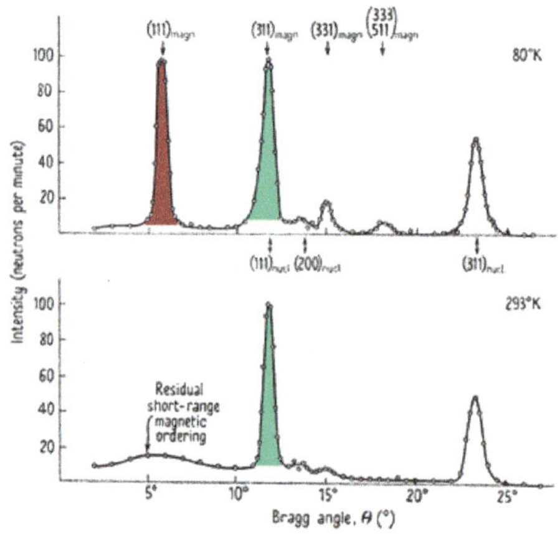

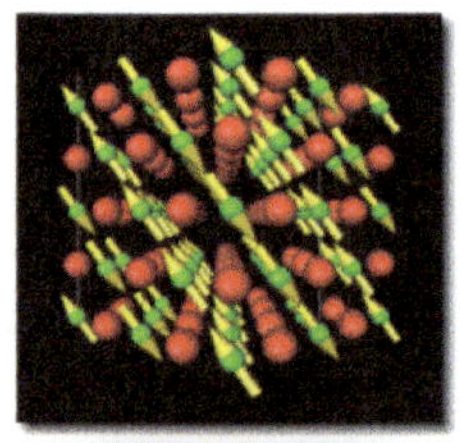

The Nobel Prize in Physics 1994

The investigation of magnetic structure in MnO crystal in 1949 by Shull has triggered the scientific community to use the neutron as source for detecting magnetic structure (The magnetic moments of some atoms point up and some point down).

Understanding the Magnetic Structure

MnF2 adopts a tetragonal crystal structure (a = b > c, all angles as 90°) as shown in figure.

Only Mn carries a moment.

Magnetic structure: corner Mn shows up and body-centered Mn shows down moment, resulting an antiferromagnet (AFM).

Since AFM, TN is 67 K, easy to study.

Mn moment is 5 μB so magnetic scattering is considerably strong and hence the magnetic signal is large.

If one considers a ferromagnetic form with the Mn moments along c, all of Mn moments point the same way, so they are equivalent, and the magnetic scattering comes from a bct arrangement of moments.

Kittel tells us that the structure factor for bcc (or bct) requires h+k+l = 2n.

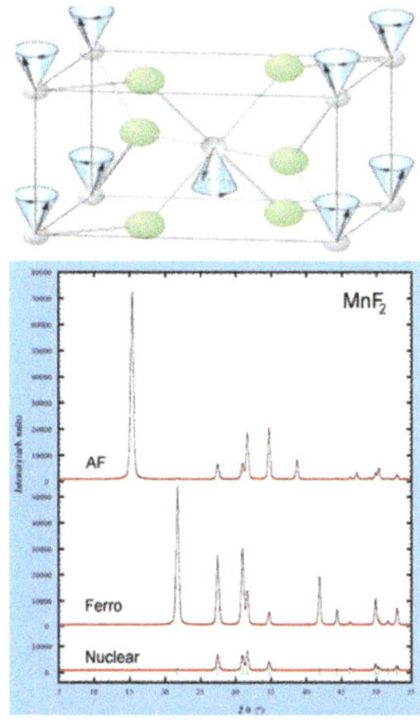

Crystal strcture and neutron diffraction of Mn2.

Magnetic Scattering from MnF_2

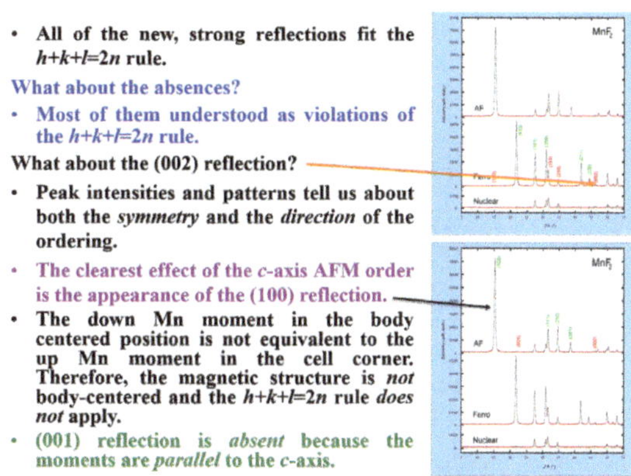

Magnetic Scattering from GdCu2

Figure depicts the typical arrangement of atoms in lattice and the moments to form different magnetic states.

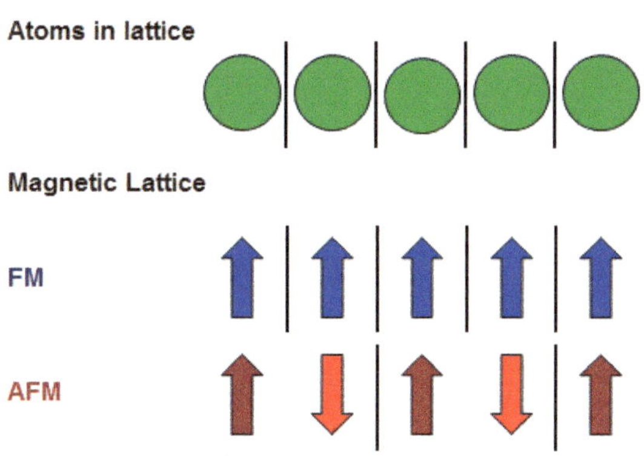

Typical arrangements of atoms in lattice and the moments.

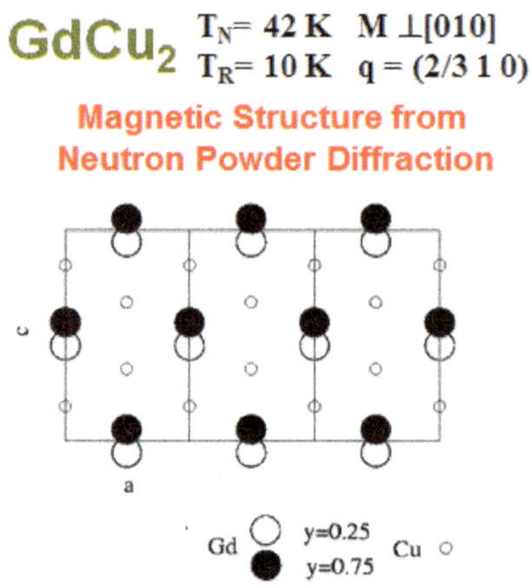

Rotter et.al. J. Magn. Mag. Mat. 214 (2000) 281

Figure shows the neutron diffraction data obtained at two different temperatures. This clearly reveals development of additional peaks at low temperature [Martin Rotter, University of Oxford].

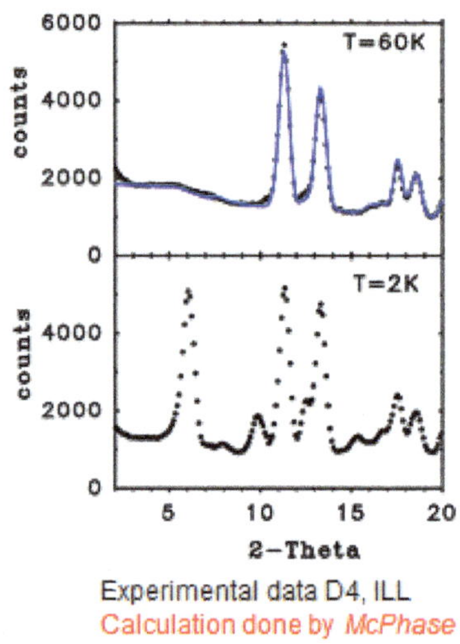

Neutron diffraction of GdCu2 data obtained at two different temperatures.

The figures demonstrate the various possible arrangements of moments and the resulting neutron diffraction patterns for GdCu2 at two different temperatures.

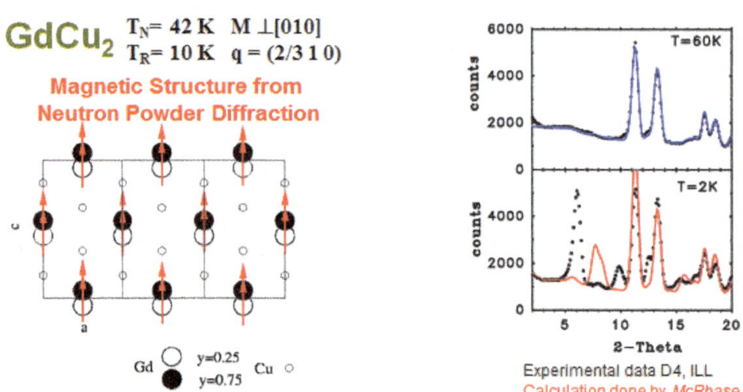

Martin Rotter, University of Oxford

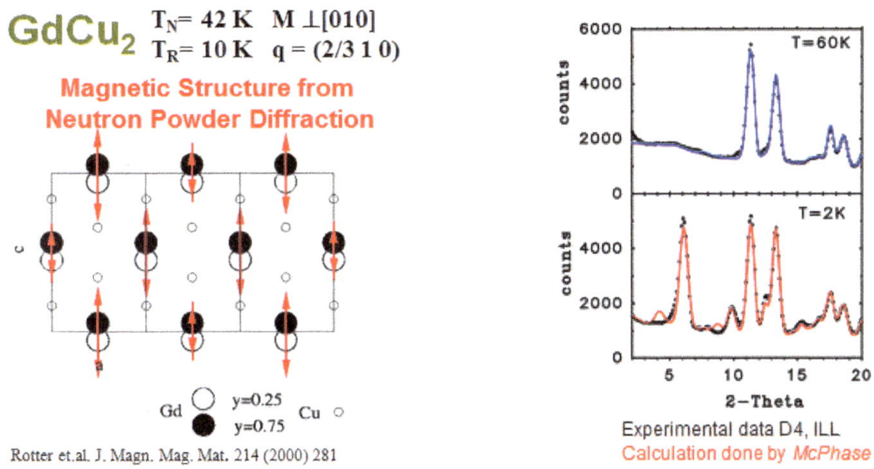

Martin Rotter, University of Oxford

Neutron diffraction patterns for $GdCu_2$ at two different temperatures with different arrangments of moments.

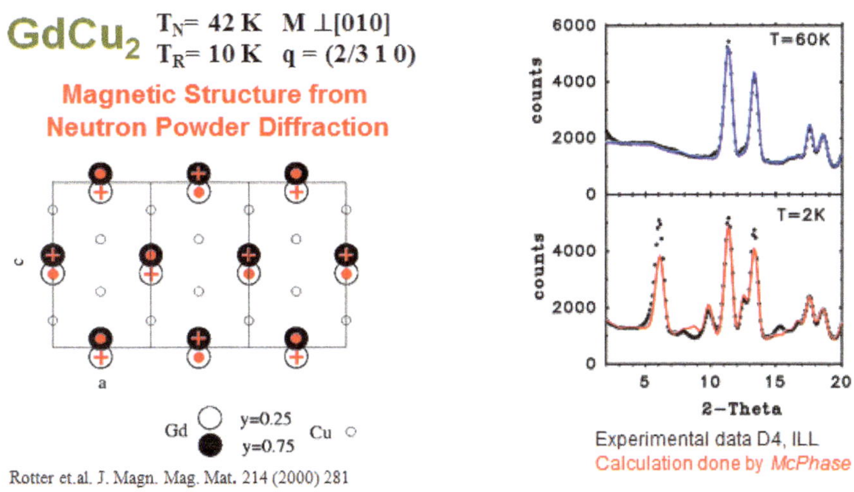

Martin Rotter, University of Oxford

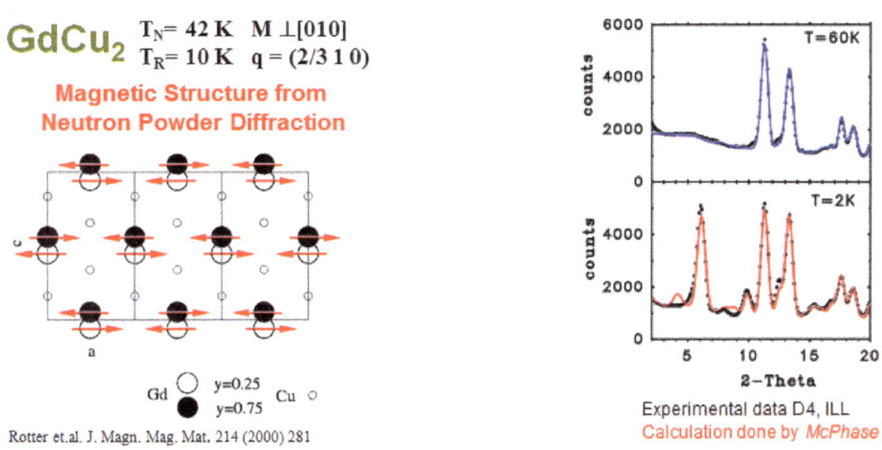

Martin Rotter, University of Oxford

Neutron diffraction patterns for GdCu2 at two different temperatures with different arrangments of moments.

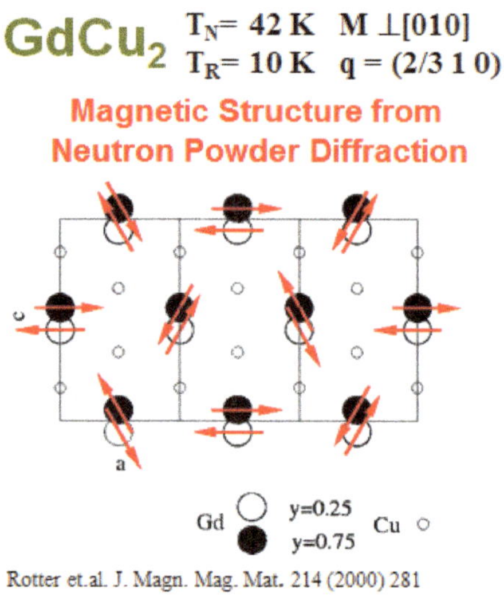

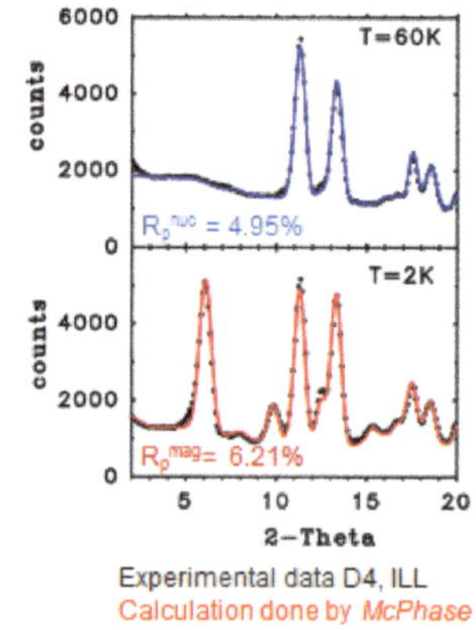

Goodness of fit

$$R_p = 100\frac{\sum_{hkl}\left|I_{calc}(hkl) - I_{exp}(hkl)\right|}{\sum_{hkl}\left|I_{exp}(hkl)\right|}$$

Some key differences:

- Magnetic scattering involves the interaction of the neutron's magnetic moment with that of an atomic electron → EM not strong interaction
- Both the nature of the scattering and the object that the neutron scatters from are different.
- The form of the scattering is therefore quite different.

Low-energy Electron Diffraction

Low-energy electron diffraction (LEED) is one of the principal techniques for the determination of surface structures. This technique can be used in one of the following two ways:

- Qualitatively:

The diffraction pattern is recorded and analysis of the spot positions yields information on the size, symmetry and rotational alignment of the adsorbate unit cell with respect to the substrate unit cell.

- Quantitatively:

The intensities of the various diffracted beams are recorded as a function of the incident electron beam energy to generate so-called I-V curves which by comparison with theoretical curves may provide accurate information on atomic positions.

LEED: Principle of Operation

A beam of electrons of a well-defined low energy (typically in the range 20-400 eV) is incident normally on the sample as shown in figure. The diffracted electrons can be observed on a fluorescent screen after energy-filtering grids, which selects only the electrons with the same kinetic energy as the primary electrons.

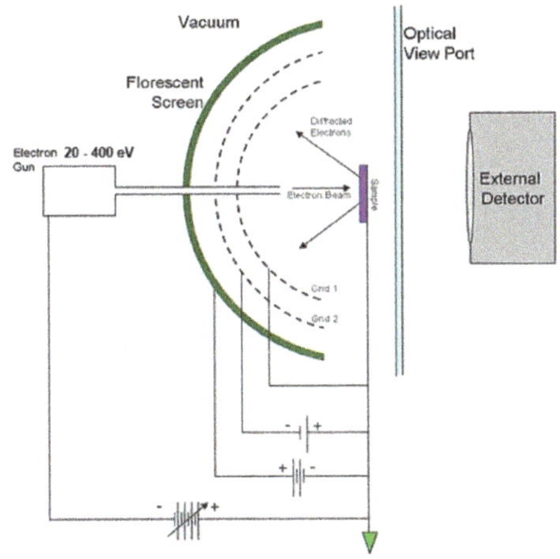

Schematic diagram of LEED set up.

LEED: Principle of Operation

By the principles of wave-particle duality, the beam of electrons may be equally regarded as a succession of electron waves incident normally on the sample. These waves will be scattered by regions of high localized electron density, i.e. the surface atoms, which can therefore be considered to act as point scatterers.

The wavelength of the electrons is given by the de-Broglie relation:

$$\lambda = \frac{h}{p}$$

where

$$p = mv = (2mE_k)^{\frac{1}{2}} = (2m.e.V)^{\frac{1}{2}} \Rightarrow \lambda = \frac{h}{(2m.e.V)^{\frac{1}{2}}}$$

m is mass of electron, v is velocity, e is electronic charge and V is acceleration voltage in eV.

The eqn. for λ provides the values of 2.74 - 0.87 Å for the energy range 20 to 200 eV. This is seen to be comparable with atomic spacing, suitable for diffraction condition associated with the atomic structure.

Let us consider a one dimensional chain of atoms as shown in figure separated by a distance a and the electron beam incident normal to the chain.

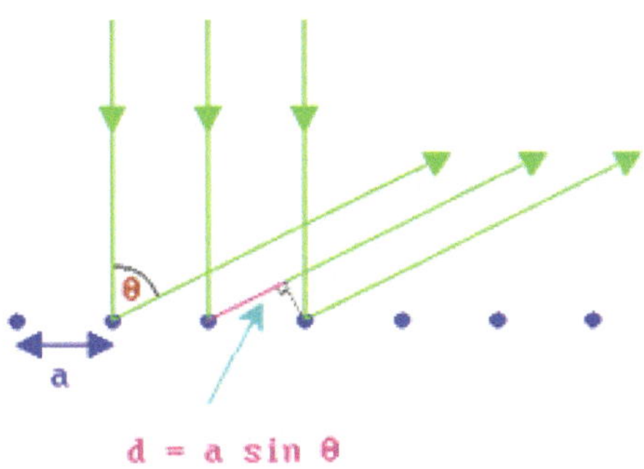

Schematic of lattice and e-beam incident.

For the constructive interference, the path difference between the wave fronts scattered by the two nearest atoms must be equal to the integral multiple of wavelengths.

$$d = a \sin\theta = n\lambda$$

where, asinθ is the path difference between two scattered waves.

Figure demonstrates the formation of diffraction pattern from the LEED.

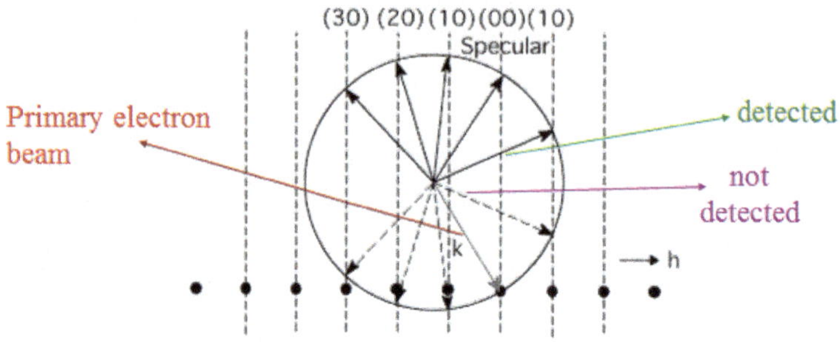

Schematic of diffraction pattern formation.

- Only a few beams get diffracted and generate diffraction patterns. This can can be seen clealy from the 2D reciprocal lattice.
- Some beams scattered over the surface (invisible)
- Some beams back-scattered towards the source
- More beams visible as k increases (smaller λ, larger K.E.)

The typical intensity profile obtained for this case is shown in figure.

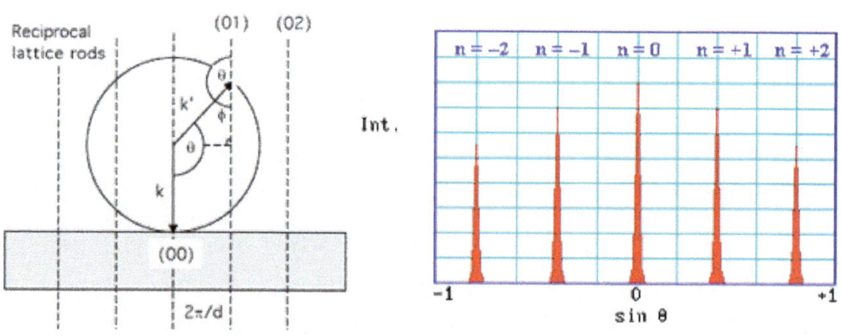

Intensity profile pattern with the angle, θ.

(1) the pattern is symmetric about $\theta = 0$ (or $\sin\theta = 0$)

(2) Since $\lambda \propto 1/(V^{1/2})$, $\sin\theta \propto 1/(V^{1/2})$

(3) sinθ is also inversely proportional to the lattice parameter, a

This reveals that all surface diffraction patterns show a symmetry reflecting that of the surface structure and of a scale showing an inverse relationship to both the square root of the electron energy and the size of the surface unit cell.

The Energy of the Electrons in LEED

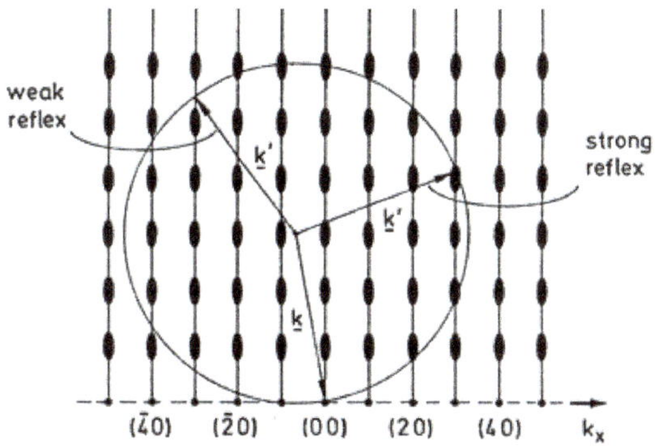

Energy of electron and diffraction pattern.

The short free path length of electrons with 20 - 200 eV for LEED gives very high surface sensitivity. However, the electrons still penetrate several atomic layers. The scattering from the several layers modulates the 2D Ewald's construction toward 3D case as shown in figure. This gives the variation in intensity of each diffracted spot with changing electron energy.

Pattern Analysis

The arrangement of atoms depends on how the crystal is cut.

Figure shows the typical example of schematic representation of three different surfaces of a face-centered cubic (fcc) metal.

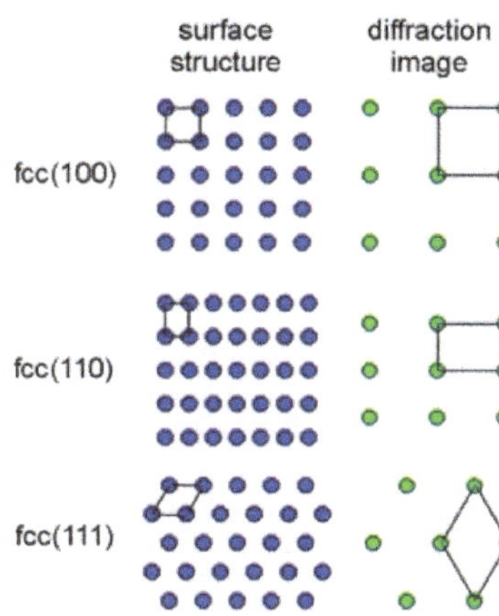

Schematic of surface structure and diffraction image for fcc structure.

The black polygons show the unit cell (smallest unit of surface) and reveal how the diffraction image appears. One important occurrence observed is that the long side of the 110 surface unit cells transform to the short side of the unit cell in the diffraction image.

The typical disadvantage of LEED is that close to normal incidence of the beam is necessary. Therefore, one can not control the surface geometry, i.e., during film growth (epitaxy). For this purpose RHEED is used.

Low-energy electron diffraction (LEED) is a technique for the determination of the surface structure of single-crystalline materials by bombardment with a collimated beam of low energy electrons (20–200 eV) and observation of diffracted electrons as spots on a fluorescent screen.

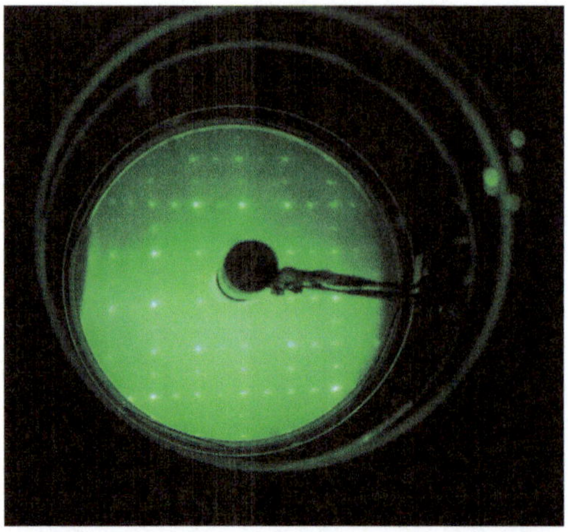

LEED pattern of a Si(100) reconstructed surface. The underlying lattice is a square lattice while the surface reconstruction has a 2x1 periodicity. As discussed in the text, the pattern shows that reconstruction exists in symmetrically equivalent domains which are oriented along different crystallographic axes. The diffraction spots are generated by acceleration of elastically scattered electrons onto a hemispherical fluorescent screen. Also seen is the electron gun which generates the primary electron beam. It covers up parts of the screen.

LEED may be used in one of two ways:

1. Qualitatively, where the diffraction pattern is recorded and analysis of the spot positions gives information on the symmetry of the surface structure. In the presence of an adsorbate the qualitative analysis may reveal information about the size and rotational alignment of the adsorbate unit cell with respect to the substrate unit cell.

2. Quantitatively, where the intensities of diffracted beams are recorded as a function of incident electron beam energy to generate the so-called I-V curves. By comparison with theoretical curves, these may provide accurate information on atomic positions on the surface at hand.

Historical Perspective

Davisson and Germer's Discovery of Electron Diffraction

The theoretical possibility of the occurrence of electron diffraction first emerged in 1924 when Louis de Broglie introduced wave mechanics and proposed the wavelike nature of all particles. In his Nobel laureated work de Broglie postulated that the wavelength of a particle with linear momentum *p* is given by *h/p*, where *h* is Planck's constant. The de Broglie hypothesis was confirmed experimentally at Bell Labs in 1927 when Clinton Davisson and Lester Germer fired low-energy electrons at a crystalline nickel target and observed that the angular dependence of the intensity of backscattered electrons showed diffraction patterns. These observations were consistent with the diffraction theory for X-rays developed by Bragg and Laue earlier. Before the acceptance of the de Broglie hypothesis diffraction was believed to be an exclusive property of waves.

Davisson and Germer published notes of their electron diffraction experiment result in Nature and in Physical Review in 1927. One month after Davisson and Germer's work appeared, Thompson and Reid published their electron diffraction work with higher kinetic energy (thousand times higher than the energy used by Davisson and Germer) in the same journal. Those experiments revealed the wave property of electrons and opened up an era of electron diffraction study.

Development of LEED as a Tool in Surface Science

Though discovered in 1927, low energy electron diffraction did not become a popular tool for surface analysis until the early 1960s. The main reasons were that monitoring directions and intensities of diffracted beams was a difficult experimental process due to inadequate vacuum techniques and slow detection methods such as a Faraday cup. Also, since LEED is a surface sensitive method, it required well-ordered surface structures. Techniques for the reconstruction of clean metal surfaces first became available much later. In the early 1960s LEED experienced a renaissance as ultra high vacuum became widely available and the post acceleration detection method was introduced. Using this technique diffracted electrons were accelerated to high energies to produce clear and visible diffraction patterns on a fluorescent screen.

It soon became clear that the kinematic (single scattering) theory, which had been successfully used to explain X-ray diffraction experiments, was inadequate for the quantitative interpretation of experimental data obtained from LEED. At this stage a detailed determination of surface structures, including adsorption sites, bond angles and bond lengths was not possible. A dynamical electron diffraction theory which took into account the possibility of multiple scattering was established in the late 1960s. With this theory it later became possible to reproduce experimental data with high precision.

Experimental Setup

In order to keep the studied sample clean and free from unwanted adsorbates, LEED experiments are performed in an ultra-high-vacuum environment (10^{-9} mbar).

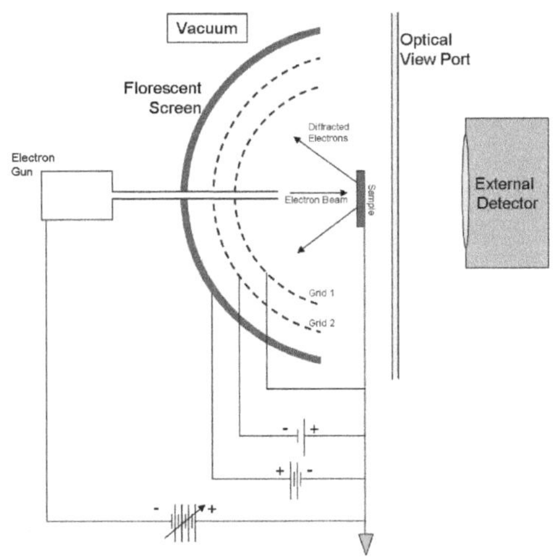

Diagram of a LEED optics apparatus.

The most important elements in a LEED experiment are:

1. A sample holder with the prepared sample
2. An electron gun
3. A display system, usually a hemispherical fluorescent screen on which the diffraction pattern can be observed directly

4. A sputtering gun for cleaning the surface
5. An Auger electron spectroscopy system in order to determine the purity of the surface.

A simplified sketch of an LEED setup is shown in figure.

Sample Preparation

The sample is usually prepared outside the vacuum chamber by cutting a slice of around 1 mm in thickness and 1 cm in diameter along the desired crystallographic axis. The correct alignment of the crystal can be achieved with the help of x-ray methods and should be within 1° of the desired angle. After being mounted in the UHV chamber the sample is chemically cleaned and flattened. Unwanted surface contaminants are removed by ion sputtering or by chemical processes such as oxidation and reduction cycles. The surface is flattened by annealing at high temperatures. Once a clean and well-defined surface is prepared, monolayers can be adsorbed on the surface by exposing it to a gas consisting of the desired adsorbate atoms or molecules.

Often the annealing process will let bulk impurities diffuse to the surface and therefore give rise to a re-contamination after each cleaning cycle. The problem is that impurities which adsorb without changing the basic symmetry of the surface, cannot easily be identified in the diffraction pattern. Therefore, in many LEED experiments Auger electron spectroscopy is used to accurately determine the purity of the sample.

Electron Gun

In the electron gun, monochromatic electrons are emitted by a cathode filament which is at a negative potential, typically 10-600 V, with respect to the sample. The electrons are accelerated and focused into a beam, typically about 0.1 to 0.5 mm wide, by a series of electrodes serving as electron lenses. Some of the electrons incident on the sample surface are backscattered elastically, and diffraction can be detected if sufficient order exists on the surface. This typically requires a region of single crystal surface as wide as the electron beam, although sometimes polycrystalline surfaces such as highly oriented pyrolytic graphite (HOPG) are sufficient.

Detector System

A LEED detector usually contains three or four hemispherical concentric grids and a phosphor screen or other position-sensitive detector. The grids are used for screening out the inelastically scattered electrons. Most new LEED systems use a reverse view scheme, which has a minimized electron gun, and the pattern is viewed from behind through a transmission screen and a viewport. Recently, a new digitized position sensitive detector called a delay-line detector with better dynamic range and resolution has been developed.

The LEED contains a retarding field analyzer to block inelastically scattered electrons. Because only spherical fields around the sampled point are allowed and the geometry of the sample and the surrounding area is not spherical, no field is allowed. Therefore, the first grid screens separates the space above the sample from the retarding field. The next grid is at a potential to block low energy electrons, it is called the suppressor or the gate. To make the retarding field homogeneous and mechanically more stable this grid often consists of two grids. The fourth grid is only necessary when the LEED is used like a tetrode and the current at the screen is measured, when it serves as screen between the gate and the anode.

Using the Detector for Auger Electron Spectroscopy

To improve the measured signal in Auger electron spectroscopy, the gate voltage is scanned in a linear ramp. An RC circuit serves to derive the second derivative, which is then amplified and digitized. To reduce the noise, multiple passes are summed up. The first derivative is very large due to the residual capacitive coupling between gate and the anode and may degrade the performance of the circuit. By applying a negative ramp to the screen this can be compensated. It is also possible to add a small sine to the gate. A high Q RLC circuit is tuned to the second harmonic to detect the second derivative.

Data Acquisition

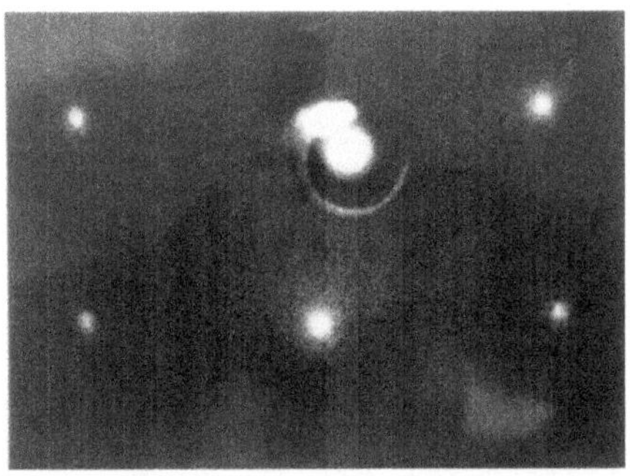

LEED pattern of a clean platinum-rhodium (100) (Miller-index) single crystal.
Taken in high vacuum using an electron gun with an energy of 85 eV.

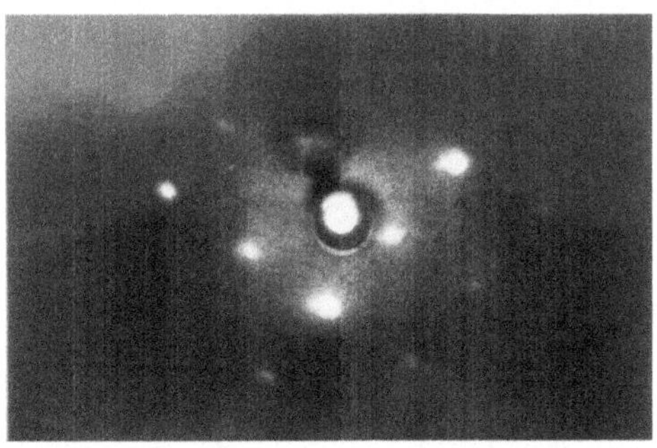

LEED pattern of CO on platinum-rhodium (100) (Miller-index) surface of a single crystal.
Taken in high vacuum using an electron gun with an energy of 94 eV.

A modern data acquisition system usually contains a CCD/CMOS camera pointed to the screen for diffraction pattern visualization and a computer for data recording and further analysis.

The shown images are examples of LEED diffraction patterns. The difference between image 1 and 2 is remarkable; where image 1 is of a clean (100) platinum-rhodium single crystal, and image 2 of the same crystal with CO adsorbed on the surface. The original surface order of the clean crystal is clearly visible in image 1, it shows a C(1x1) structure; the extra spots in image 2 are caused by the CO on the surface and are an example of a C(2x2) structure. The diffraction spots are generated by acceleration of elastically scattered electrons onto a hemispherical fluorescent screen, a retarding field analyzer. In the middle one can see the bright spot of the electron gun which generates the primary electron beam.

Theory

Surface Sensitivity

The basic reason for the high surface sensitivity of LEED is that for low-energy electrons the interaction between the solid and electrons is especially strong. Upon penetrating the crystal, primary electrons will lose kinetic energy due to inelastic scattering processes such as plasmon- and phonon excitations as well as electron-electron interactions. In cases where the detailed nature of the inelastic processes is unimportant they are commonly treated by assuming an exponential decay of the primary electron beam intensity, I_0, in the direction of propagation:

$$I(d) = I_0 * e^{-d/\Lambda(E)}$$

Here d is the penetration depth and $\Lambda(E)$ denotes the inelastic mean free path, defined as the distance an electron can travel before its intensity has decreased by the factor $1/e$. While the inelastic scattering processes and consequently the electronic mean free path depend on the energy, it is relatively independent of the material. The mean free path turns out to be minimal (5–10 Å) in the energy range of low-energy electrons (20–200 eV). This effective attenuation means that only a few atomic layers are sampled by the electron beam and as a consequence the contribution of deeper atoms to the diffraction progressively decreases.

Kinematic Theory: Single Scattering

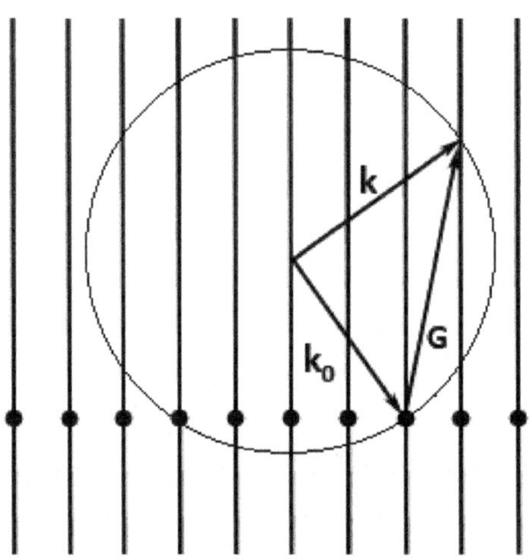

Ewald's sphere construction for the case of diffraction from a 2D-lattice. The intersections between Ewald's sphere and reciprocal lattice rods define the allowed diffracted beams.

Kinematic diffraction is defined as the situation where electrons impinging on a well-ordered crystal surface are elastically scattered only once by that surface. In the theory the electron beam is represented by a plane wave with a wavelength in accordance to the de Broglie hypothesis:

$$\lambda = \frac{h}{\sqrt{2mE}}, \qquad \lambda[\text{nm}] \approx \sqrt{\frac{1.5}{E[\text{eV}]}}$$

The interaction between the scatterers present in the surface and the incident electrons is most conveniently described in reciprocal space. In three dimensions the primitive reciprocal lattice vectors are related to the real space lattice {a, b, c} in the following way:

$$\mathbf{a}^* = \frac{2\pi \mathbf{b} \times \mathbf{c}}{\mathbf{a} \cdot (\mathbf{b} \times \mathbf{c})},$$

$$\mathbf{b}^* = \frac{2\pi \mathbf{c} \times \mathbf{a}}{\mathbf{b} \cdot (\mathbf{c} \times \mathbf{a})},$$

$$\mathbf{c}^* = \frac{2\pi \mathbf{a} \times \mathbf{b}}{\mathbf{c} \cdot (\mathbf{a} \times \mathbf{b})}$$

For an incident electron with wave vector $\mathbf{k}_0 = 2\pi / \lambda_0$ and scattered wave vector $\mathbf{k} = 2\pi / \lambda$, the condition for constructive interference and hence diffraction of scattered electron waves is given by the Laue condition

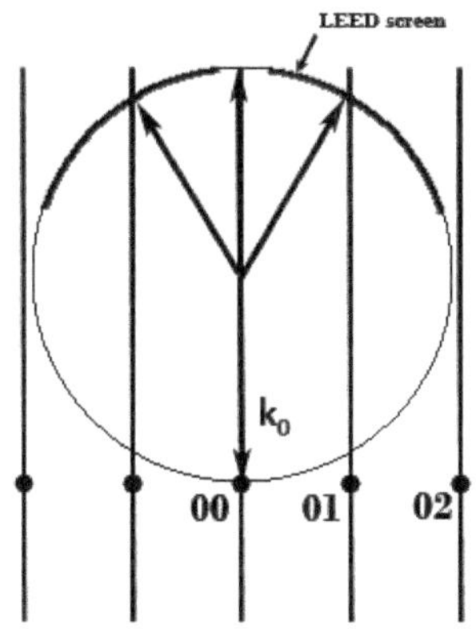

Ewald's sphere construction for the case of normal incidence of the primary electron beam. The diffracted beams are indexed according to the values of *h* and *l*.

$$\mathbf{k} - \mathbf{k}_0 = \mathbf{G}_{hkl}, (1)$$

where (*h*,*k*,*l*) is a set of integers and

$$\mathbf{G}_{hkl} = h\mathbf{a}^* + k\mathbf{b}^* + l\mathbf{c}^*$$

is a vector of the reciprocal lattice. The magnitudes of the wave vectors are unchanged, i.e. $|\mathbf{k}_0| = |\mathbf{k}|$, since only elastic scattering is considered. Since the mean free path of low energy electrons in a crystal is only a few angstroms, only the first few atomic layers contribute to the diffraction. This means that there are no diffraction conditions in the direction perpendicular to the sample surface. As a consequence the reciprocal lattice of a surface is a 2D lattice with rods extending perpendicular from each lattice point. The rods can be pictured as regions where the reciprocal lattice points are infinitely dense. Therefore, in the case of diffraction from a surface equation (1) reduces to the 2D form:

$$\mathbf{k}^{\parallel} - \mathbf{k}_0^{\parallel} = \mathbf{G}_{hk} = h\mathbf{a}^* + k\mathbf{b}^*, (2)$$

where $\mathbf{a}^*$ and $\mathbf{b}^*$ are the primitive translation vectors of the 2D reciprocal lattice of the surface and $\mathbf{k}^{\parallel}, \mathbf{k}_0^{\parallel}$ denote the component of respectively the reflected and incident wave vector parallel to the sample surface. $\mathbf{a}^*$ and $\mathbf{b}^*$ are related to the real space surface lattice in the following way:

$$\mathbf{a}^* = \frac{2\pi \mathbf{b} \times \hat{\mathbf{n}}}{|\mathbf{a} \times \mathbf{b}|}$$

$$\mathbf{b}^* = \frac{2\pi \hat{\mathbf{n}} \times \mathbf{a}}{|\mathbf{a} \times \mathbf{b}|}$$

The Laue condition equation (2) can readily be visualized using the Ewald's sphere construction. Figure shows a simple illustration of this principle: The wave vector k_0 of the incident electron beam is drawn such that it terminates at a reciprocal lattice point. The Ewald's sphere is then the sphere with radius $|k_0|$ and origin at the center of the incident wave vector.

By construction, every wave vector centered at the origin and terminating at an intersection between a rod and the sphere will then satisfy the Laue condition and thus represent an allowed diffracted beam.

Interpretation of LEED Patterns

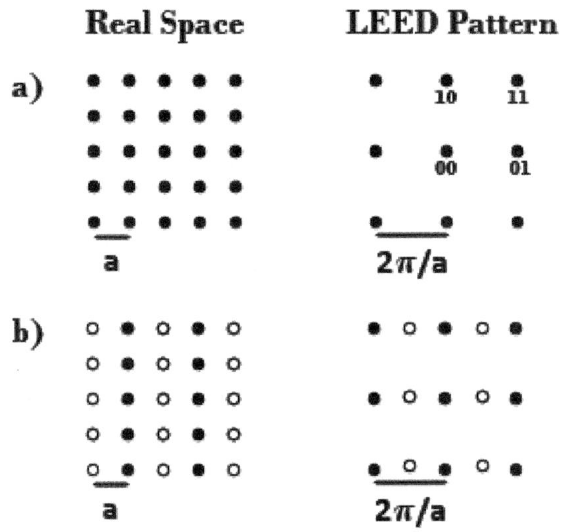

Real space- and reciprocal lattices for the case of **a)** a (100) face of a simple cubic lattice and **b)** a (2×1) commensurate superstructure. The white spots in the LEED pattern are the *extra spots* associated with the adsorbate structure.

Figure shows the Ewald's sphere for the case of normal incidence of the primary electron beam, as would be the case in an actual LEED setup. It is apparent that the pattern observed on the fluorescent screen is a direct picture of the reciprocal lattice of the surface. The size of the Ewald's sphere and hence the number of diffraction spots on the screen is controlled by the incident electron energy. From the knowledge of the reciprocal lattice models for the real space lattice can be constructed and the surface can be characterized at least qualitatively in terms of the surface periodicity and the point group. Figure a shows a model of an unreconstructed (100) face of a simple cubic crystal and the expected LEED pattern. The spots are indexed according to the values of *h* and *k*.

Superstructures

We now consider the case of an overlaying superstructure on a substrate surface. If the LEED pattern of the underlying (1×1) surface is known, spots due to the superstructure can be identified as *extra spots* or *super spots*. Figure b shows the simple example of a (2×1) superstructure on a square lattice.

For a commensurate superstructure the symmetry and the rotational alignment with respect to adsorbent surface can be determined from the LEED pattern. This is easiest shown by using a matrix notation, where the primitive translation vectors of the superlattice $\{a_s, b_s\}$ are linked to the primitive translation vectors of the underlying (1x1) lattice {a,b} in the following way:

$$\mathbf{a}_s = G_{11}\mathbf{a} + G_{12}\mathbf{b},$$
$$\mathbf{b}_s = G_{21}\mathbf{a} + G_{22}\mathbf{b}.$$

The matrix for the superstructure then is

$$G = \begin{pmatrix} G_{11} & G_{12} \\ G_{21} & G_{22} \end{pmatrix}.$$

Similarly, the primitive translation vectors of the lattice describing the *extra spots* $\{a_s^*, b_s^*\}$ are linked to the primitive translation vectors of the reciprocal lattice $\{a^*, b^*\}$

$$\mathbf{a}_s^* = G_{11}^*\mathbf{a}^* + G_{12}^*\mathbf{b}^*,$$
$$\mathbf{b}_s^* = G_{21}^*\mathbf{a}^* + G_{22}^*\mathbf{b}^*.$$

G^* is related to G in the following way

$$G^* = (G^{-1})^T$$
$$= \frac{1}{\det(G)} \begin{pmatrix} G_{22} & -G_{21} \\ -G_{12} & G_{11} \end{pmatrix}.$$

Domains

An essential problem when considering LEED patterns is the existence of symmetrically equivalent domains. Domains may lead to diffraction patterns which have higher symmetry than the actual surface at hand. The reason is that usually the cross sectional area of the primary electron beam (~1 mm²) is large compared to the average domain size on the surface and hence the LEED pattern might be a superposition of diffraction beams from domains oriented along different axes of the substrate lattice.

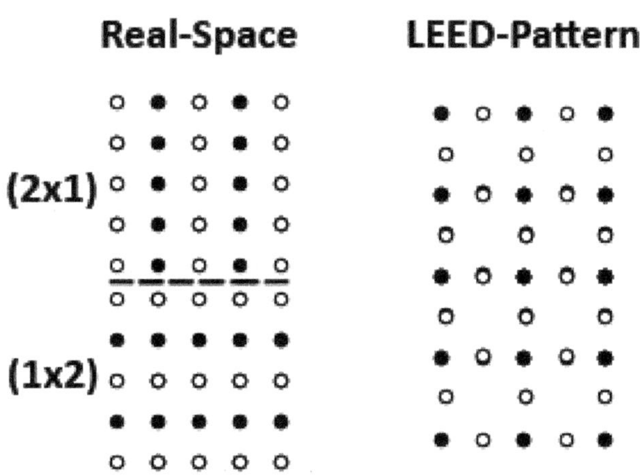

Superposition of the LEED patterns associated with the two orthogonal domains (2x1) and (1x2). The LEED pattern has a fourfold rotational symmetry.

However, since the average domain size generally is larger than the coherence length of the probing electrons, interference between electrons scattered from different domains can be neglected. Therefore, the total LEED pattern emerges as the incoherent sum of the diffraction patterns associated with the individual domains.

Figure shows the superposition of the diffraction patterns for the two orthogonal domains (2x1) and (1x2) on a square lattice, i.e. for the case where one structure is just rotated by 90° with respect to the other. The (2x1) structure and the respective LEED pattern are shown in figure. It is apparent that the local symmetry of the surface structure is twofold while the LEED pattern exhibits a fourfold symmetry.

Figure shows a real diffraction pattern of the same situation for the case of a Si(100) surface. However, here the (2x1) structure is formed due to surface reconstruction.

Dynamical Theory: Multiple Scattering

The inspection of the LEED pattern gives a qualitative picture of the surface periodicity i.e. the size of the surface unit cell and to a certain degree of surface symmetries. However it will give no information about the atomic arrangement within a surface unit cell or the sites of adsorbed atoms. For instance if the whole superstructure in figure b is shifted such that the atoms adsorb in bridge sites instead of on-top sites the LEED pattern will be the same.

A more quantitative analysis of LEED experimental data can be achieved by analysis of so-called I-V curves, which are measurements of the intensity versus incident electron energy. The I-V curves can be recorded by using a camera connected to computer controlled data handling or by direct measurement with a movable Faraday cup. The experimental curves are then compared to computer calculations based on the assumption of a particular model system. The model is changed in an iterative process until a satisfactory agreement between experimental and theoretical curves is achieved. A quantitative measure for this agreement is the so-called *reliability*- or R-factor. A commonly used reliability factor is the one proposed by Pendry. It is expressed in terms of the logarithmic derivative of the intensity:

$$L(E) = I' / I.$$

The R-factor is then given by:

$$R = \sum_g \int (Y_{gth} - Y_{gexpt})^2 dE / \sum_g \int (Y_{gth}^2 + Y_{gexpt}^2) dE,$$

where $Y(E) = L^{-1} / (L^{-2} + V_{oi}^2)$ and V_{oi} is the imaginary part of the electron self-energy. In general, $R_p \leq 0.2$ is considered as a good agreement, $R_p \simeq 0.3$ is considered mediocre and $R_p \simeq 0.5$ is considered a bad agreement. Figure shows examples of the comparison between experimental I-V spectra and theoretical calculations.

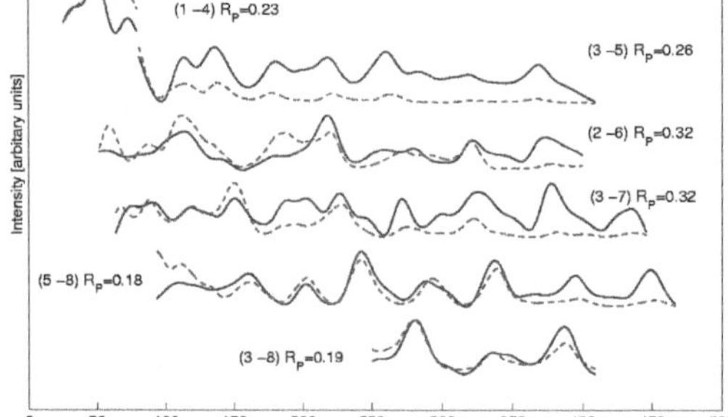

Examples of the comparison between experimental data and a theoretical calculation (an AlNiCo quasicrystal surface). Thanks to R. Diehl and N. Ferralis for providing the data.

Dynamical LEED Calculations

The term *dynamical* stems from the studies of X-ray diffraction and describes the situation where the response of the crystal to an incident wave is included self-consistently and multiple scattering can occur. The aim of any dynamical LEED theory is to calculate the intensities of diffraction of an electron beam impinging on a surface as accurately as possible.

A common method to achieve this is the self-consistent multiple scattering approach. One essential point in this approach is the assumption that the scattering properties of the surface, i.e. of the individual atoms, are known in detail. The main task then reduces to the determination of the effective wave field incident on the individual scatters present in the surface, where the effective field is the sum of the primary field and the field emitted from all the other atoms. This must be done in a self-consistent way, since the emitted field of an atom depends on the incident effective field upon it. Once the effective field incident on each atom is determined, the total field emitted from all atoms can be found and its asymptotic value far from the crystal then gives the desired intensities.

A common approach in LEED calculations is to describe the scattering potential of the crystal by a "muffin tin" model, where the crystal potential can be imagined being divided up by non-overlapping spheres centered at each atom such that the potential has a spherically symmetric form inside the spheres and is constant everywhere else. The choice of this potential reduces the problem to scattering from spherical potentials, which can be dealt with effectively. The task is then to solve the Schrödinger equation for an incident electron wave in that "muffin tin" potential.

Related Techniques

Tensor LEED

In LEED the exact atomic configuration of a surface is determined by a trial and error process where measured I-V curves are compared to computer-calculated spectra under the assumption of a model structure. From an initial reference structure a set of trial structures is created by varying the model parameters. The parameters are changed until an optimal agreement between theory and experiment is achieved. However, for each trial structure a full LEED calculation with multiple scattering corrections must be conducted. For systems with a large parameter space the need for computational time might become significant. This is the case for complex surfaces structures or when considering large molecules as adsorbates.

Tensor LEED is an attempt to reduce the computational effort needed by avoiding full LEED calculations for each trial structure. The scheme is as follows: One first defines a reference surface structure for which the I-V spectrum is calculated. Next a trial structure is created by displacing some of the atoms. If the displacements are small the trial structure can be considered as a small perturbation of the reference structure and first-order perturbation theory can be used to determine the I-V curves of a large set of trial structures.

Spot Profile Analysis Low-energy Electron Diffraction

A real surface is not perfectly periodic but has many imperfections in the form of dislocations, atomic steps, terraces and the presence of unwanted adsorbed atoms. This departure from a perfect surface leads to a broadening of the diffraction spots and adds to the background intensity in the LEED pattern.

SPA-LEED is a technique where the intensity of diffraction beams is measured in order to determine the diffraction spot profiles. The spots are sensitive to the irregularities in the surface structure and their examination therefore permits more-detailed conclusions about some surface characteristics. Using SPA-LEED may for instance permit a quantitative determination of the surface roughness, terrace sizes or surface steps.

Reflection High-energy Electron Diffraction

Reflection high-energy electron diffraction (RHEED) is a technique used to characterize the surface of crystalline materials. RHEED systems gather information only from the surface layer of the sample, which distinguishes RHEED from other materials characterization methods that also rely on diffraction of high-energy electrons. Transmission electron microscopy, another common electron diffraction method samples the bulk of the sample due to the geometry of the system. Low-energy electron diffraction (LEED) is also surface sensitive, but LEED achieves surface sensitivity through the use of low energy electrons.

Introduction

A RHEED system requires an electron source (gun), photoluminescent detector screen and a sample with a clean surface, although modern RHEED systems have additional parts to optimize the technique. The electron gun generates a beam of electrons which strike the sample at a very small angle relative to the sample surface. Incident electrons diffract from atoms at the surface of the sample, and a small fraction of the diffracted electrons interfere constructively at specific angles and form regular patterns on the detector. The electrons interfere according to the position of atoms on the sample surface, so the diffraction pattern at the detector is a function of the sample surface. Figure shows the most basic setup of a RHEED system.

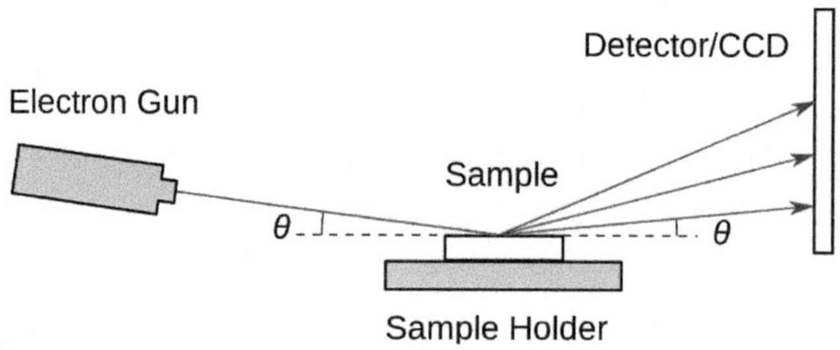

Systematic setup of the electron gun, sample and detector /CCD components of a RHEED system. Electrons follow the path indicated by the arrow and approach the sample at angle θ. The sample surface diffracts electrons, and some of these diffracted electrons reach the detector and form the RHEED pattern. The reflected (specular) beam follows the path from the sample to the detector.

Surface Diffraction

In the RHEED setup, only atoms at the sample surface contribute to the RHEED pattern. The glancing angle of incident electrons allows them to escape the bulk of the sample and to reach the detector. Atoms at the sample surface diffract (scatter) the incident electrons due to the wavelike properties of electrons.

The diffracted electrons interfere constructively at specific angles according to the crystal structure and spacing of the atoms at the sample surface and the wavelength of the incident electrons. Some of the electron waves created by constructive interference collide with the detector, creating specific diffraction patterns according to the surface features of the sample. Users characterize the crystallography of the sample surface through analysis of the diffraction patterns. Figure shows a RHEED pattern.

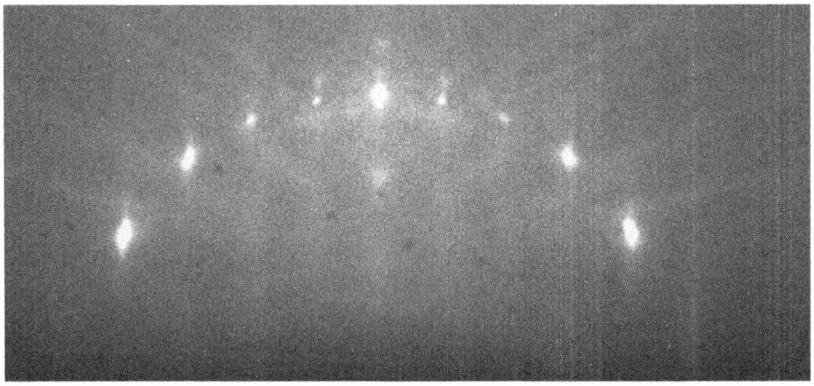

A RHEED pattern obtained from electron diffraction from a clean TiO2 (110) surface. The bright spots indicate where many electrons reach the detector. The lines that can be observed are Kikuchi Lines.

Two types of diffraction contribute to RHEED patterns. Some incident electrons undergo a single, elastic scattering event at the crystal surface, a process termed kinematic scattering. Dynamic scattering occurs when electrons undergo multiple diffraction events in the crystal and lose some of their energy due to interactions with the sample. Users extract non-qualitative data from the kinematically diffracted electrons. These electrons account for the high intensity spots or rings common to RHEED patterns. RHEED users also analyze dynamically scattered electrons with complex techniques and models to gather quantitative information from RHEED patterns.

Kinematic Scattering Analysis

RHEED users construct Ewald's spheres to find the crystallographic properties of the sample surface. Ewald's spheres show the allowed diffraction conditions for kinematically scattered electrons in a given RHEED setup. The diffraction pattern at the screen relates to the Ewald's sphere geometry, so RHEED users can directly calculate the reciprocal lattice of the sample with a RHEED pattern, the energy of the incident electrons and the distance from the detector to the sample. The user must relate the geometry and spacing of the spots of a perfect pattern to the Ewald's sphere in order to determine the reciprocal lattice of the sample surface.

The Ewald's sphere analysis is similar to that for bulk crystals, however the reciprocal lattice for the sample differs from that for a 3D material due to the surface sensitivity of the RHEED process. The reciprocal lattices of bulk crystals consist of a set of points in 3D space. However, only the first few layers of the material contribute to the diffraction in RHEED, so there are no diffraction conditions in the dimension perpendicular to the sample surface. Due to the lack of a third diffracting condition, the reciprocal lattice of a crystal surface is a series of infinite rods extending perpendicular to the sample's surface. These rods originate at the conventional 2D reciprocal lattice points of the sample's surface.

The Ewald's sphere is centered on the sample surface with a radius equal to the reciprocal of the wavelength of the incident electrons. The relationship is given by

$$k_0 = \frac{2\pi}{\lambda},$$

where λ is the wavelength of incident electrons.

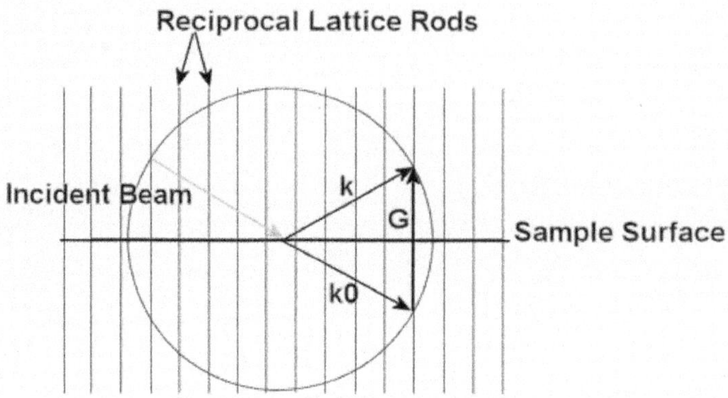

The RHEED construction of the Ewald's Sphere at the sample surface. The radius of the Ewald's sphere is equal to the magnitude of the ko vector, which is the reciprocal of the wavelength of the incident electrons. The k vector, corresponds to an allowed diffraction condition, and the G vector is the difference between the k and k_o vectors. In this figure, the k vector is the specular beam.

Diffraction conditions are satisfied where the rods of reciprocal lattice intersect the Ewald's sphere. Therefore, the magnitude of a vector from the origin of the Ewald's sphere to the intersection of any reciprocal lattice rods is equal in magnitude to that of the incident beam. Equation 2 shows this relationship.

$$|k_0| = |k_i| \quad (2)$$

Where: k_o=incident electron wave vector

k_i=electron wave vector at any intersection of reciprocal lattice with Ewald's sphere

An arbitrary vector, G, defines the reciprocal lattice vector between the ends of any two k vectors. Vector G is useful for finding distance between arbitrary planes in the crystal. Vector G is calculated using Equation 3.

$$G = k_i - k_0 \quad (3)$$

Figure shows the construction of the Ewald's sphere and provides examples of the G, k and k_o vectors.

Many of the reciprocal lattice rods meet the diffraction condition, however the RHEED system is designed such that only the low orders of diffraction are incident on the detector. The RHEED pattern at the detector is a projection only of the k vectors that are within the angular range that contains the detector. The size and position of the detector determine which of the diffracted electrons are within the angular range that reaches the detector, so the geometry of the RHEED pattern can be related back to the geometry of the reciprocal lattice of the sample surface through use of trigonometric relations and the distance from the sample to detector.

The k vectors are labeled such that the k vector that forms the smallest angle with the sample surface is called 0th order beam. The 0th order beam is also known as the specular beam. Each successive intersection of a rod and the sphere further from the sample surface is labeled as a higher order reflection. The center of the Ewald's sphere is positioned such that the specular beam forms the same angle with the substrate as the incident electron beam. The specular point has the greatest intensity on a RHEED pattern and is labeled as the (00) point by convention. The other points on the RHEED pattern are indexed according to the reflection order they project.

The radius of the Ewald's sphere is much larger than the spacing between reciprocal lattice rods because the incident beam has a very short wavelength due to its high-energy electrons. Rows of reciprocal lattice rods actually intersect the Ewald's sphere as an approximate plane because identical rows of parallel reciprocal lattice rods sit directly in front and behind the single row shown. Figure shows a cross sectional view of a single row of reciprocal lattice rods filling of the diffraction conditions. The reciprocal lattice rods in Figure show the end on view of these planes, which are perpendicular to the computer screen in the figure.

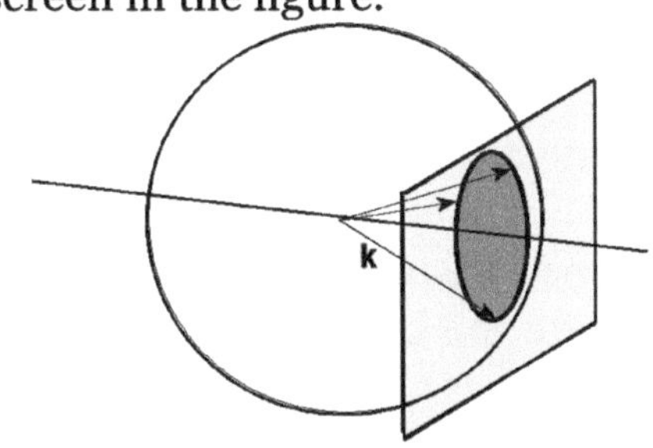

Diffraction from a row of atoms a Laue circle on the surface of the Ewald's sphere. The reciprocal lattice rods are so closely space, that they comprise the plane cutting the sphere. Diffraction conditions are fulfilled on the perimeter of the Laue circle. The vectors are all equal to the reciprocal of the incident vector, k.

The intersections of these effective planes with the Ewald's sphere forms circles, called Laue circles. The RHEED pattern is a collection of points on the perimeters of concentric Laue circles around the center point. However, interference effects between the diffracted electrons still yield strong intensities at single points on each Laue circle. Figure shows the intersection of one of these planes with the Ewald's Sphere.

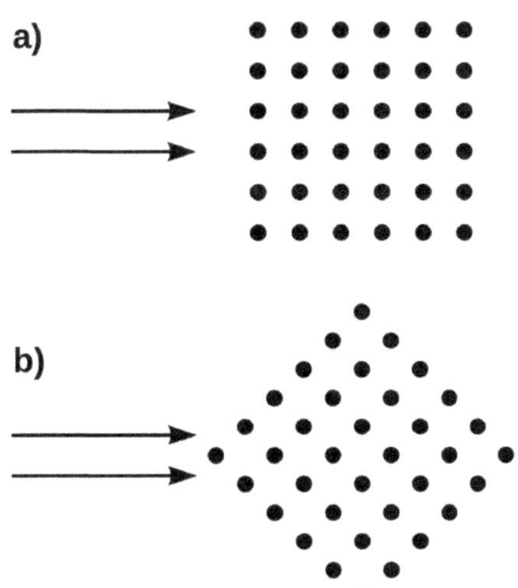

The incident electron beam is incident on an identical surface structure at a different azimuth angles in a) and b). The sample is viewed from the top in the figure, and the points correspond to the reciprocal lattice rods, which extend out of the screen. The RHEED pattern would be different for each azimuth angle.

The azimuthal angle affects the geometry and intensity of RHEED patterns. The azimuthal angle is the angle at which the incident electrons intersect the ordered crystal lattice on the surface of the sample. Most RHEED systems are equipped with a sample holder that can rotate the crystal around an axis perpendicular to the sample surface. RHEED users rotate the sample to optimize the intensity profiles of patterns. Users generally index at least 2 RHEED scans at different azimuth angles for reliable characterization of the crystal's surface structure. Figure shows a schematic diagram of an electron beam incident on the sample at different azimuth angles.

Users sometimes rotate the sample around an axis perpendicular to the sampling surface during RHEED experiments to create a RHEED pattern called the azimuthal plot. Rotating the sample changes the intensity of the diffracted beams due to their dependence on the azimuth angle. RHEED specialists characterize film morphologies by measuring the changes in beam intensity and comparing these changes to theoretical calculations, which can effectively model the dependence of the intensity of diffracted beams on the azimuth angle.

Dynamic Scattering Analysis

The dynamically, or inelastically, scattered electrons provide several types of information about the sample as well. The brightness or intensity at a point on the detector depends on dynamic scattering, so all analysis involving the intensity must account for dynamic scattering. Some inelastically scattered electrons penetrate the bulk crystal and fulfill Bragg diffraction conditions. These inelastically scattered electrons can reach the detector to yield kikuchi diffraction patterns, which are useful for calculating diffraction conditions. Kikuchi patterns are characterized by lines connecting the intense diffraction points on a RHEED pattern. Figure shows a RHEED pattern with visible Kikuchi lines.

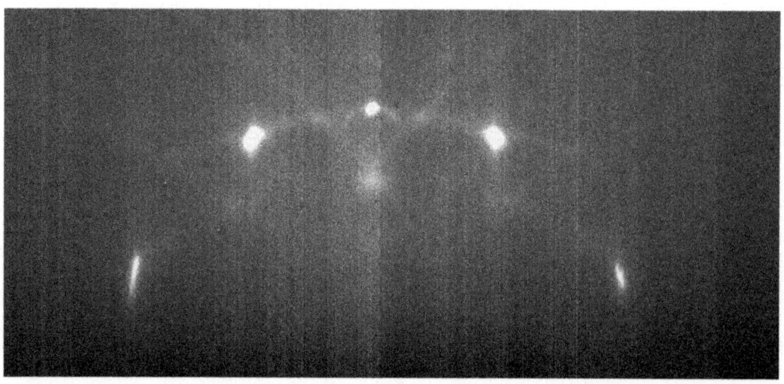

A RHEED pattern from a TiO_2 (110) surface with visible Kikuchi lines. The Kikuchi lines pass through the Laue circles and appear to radiate from the center of the pattern.

RHEED System Requirements

Electron Gun

The electron gun is one of the most important piece of equipment in a RHEED system. The gun limits the resolution and testing limits of the system. Tungsten filaments are the primary electron source for the electron gun of most RHEED systems due to the low work function of tungsten. In the typical setup, the tungsten filament is the cathode and a positively biased anode draws electrons from the tip of the tungsten filament.

The magnitude of the anode bias determines the energy of the incident electrons. The optimal anode bias is dependent upon the type of information desired. At large incident angles, electrons with high energy can penetrate the surface of the sample and degrade the surface sensitivity of the instrument. However, the dimensions of the Laue zones are proportional to the inverse square of the electron energy meaning that more information is recorded at the detector at higher incident electron energies. For general surface characterization, the electron gun is operated the range of 10-30 keV.

In a typical RHEED setup, one magnetic and one electric field focus the incident beam of electrons. A negatively biased Wehnelt electrode positioned between the cathode filament and anode applies a small electric field, which focuses the electrons as they pass through the anode. An adjustable magnetic lens focuses the electrons onto the sample surface after they pass through the anode. A typical RHEED source has a focal length around 50 cm. The beam is focused to the smallest possible point at the detector rather than the sample surface so that the diffraction pattern has the best resolution.

Phosphor screens that exhibit photoluminescence are widely used as detectors. These detectors emit green light from areas where electrons hit their surface and are common to TEM as well. The detector screen is useful for aligning the pattern to an optimal position and intensity. CCD cameras capture the patterns to allow for digital analysis.

Sample Surface

The sample surface must be extremely clean for effective RHEED experiments. Contaminants on the sample surface interfere with the electron beam and degrade the quality of the RHEED pattern. RHEED users employ two main techniques to create clean sample surfaces. Small samples can be cleaved in the vacuum chamber prior to RHEED analysis. The newly exposed, cleaved surface is analyzed. Large samples, or those that are not able to be cleaved prior to RHEED analysis can be coated with a passive oxide layer prior to analysis. Subsequent heat treatment under the vacuum of the RHEED chamber removes the oxide layer and exposes the clean sample surface.

Vacuum Requirements

Because gas molecules diffract electrons and affect the quality of the electron gun, RHEED experiments are performed under vacuum. The RHEED system must operate at a pressure low enough to prevent significant scattering of the electron beams by gas molecules in the chamber. At electron energies of 10keV, a chamber pressure of 10^{-5} mbar or lower is necessary to prevent significant scattering of electrons by the background gas. In practice, RHEED systems are operated under ultra high vacuums. The chamber pressure is minimized as much as possible in order to optimize the process. The vacuum conditions limit the types of materials and processes that can be monitored in situ with RHEED.

RHEEd Patterns of Real Surfaces

Previous analysis focused only on diffraction from a perfectly flat surface of a crystal surface. However, non-flat surfaces add additional diffraction conditions to RHEED analysis.

Streaked or elongated spots are common to RHEED patterns. As figure shows, the reciprocal lattice rods with the lowest orders intersect the Ewald sphere at very small angles, so the intersection between the rods and sphere is not a singular point if the sphere and rods have thickness. The incident electron beam diverges and electrons in the beam have a range of energies, so in practice, the Ewald sphere is not infinitely thin as it is theoretically modeled. The reciprocal lattice rods have a finite thickness as well, with their diameters dependent on the quality of the sample surface. Streaks appear in the place of perfect points when broadened rods intersect the Ewald sphere. Diffraction conditions are fulfilled over the entire intersection of the rods with the sphere, yielding elongated points or 'streaks' along the vertical axis of the RHEED pattern. In real cases, streaky RHEED patterns indicate a flat sample surface while the broadening of the streaks indicate small area of coherence on the surface.

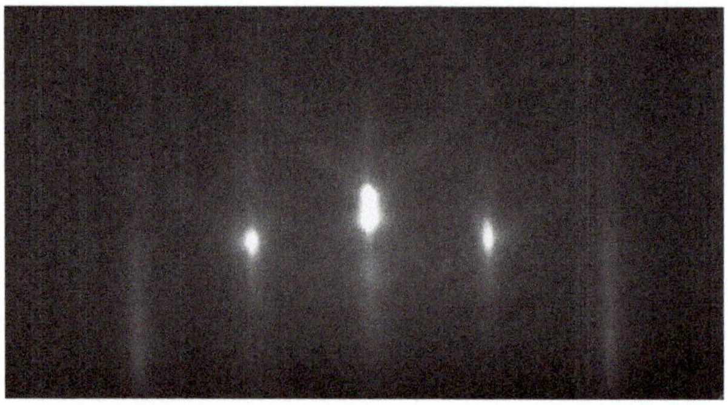

Streaked RHEED pattern from the TiO_2 (110) surface. The sample had a terraced surface, which caused noticeable streaking compared to the RHEED pattern from the flat TiO_2 (110) surface shown above.

Surface features and polycrystalline surfaces add complexity or change RHEED patterns from those from perfectly flat surfaces. Growing films, nucleating particles, crystal twinning, grains of varying size and adsorbed species add complicated diffraction conditions to those of a perfect surface. Superimposed patterns of the substrate and heterogeneous materials, complex interference patterns and degradation of the resolution are characteristic of complex surfaces or those partially covered with heterogeneous materials.

Specialized RHEED Techniques

Film Growth

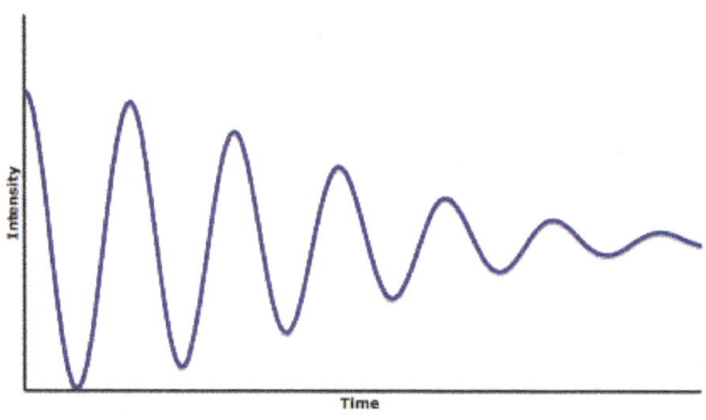

The curve is a rough model of the fluctuation of the intensity of a single RHEED point during MBE deposition. Each peak represents the forming of a new monolayer. Since the degree of order is at a maximum once a new monolayer has been formed, the spots in the diffraction pattern have maximum intensity since the maximum number of diffraction centers of the new layer contribute to the diffracted beam. The overall intensity of the oscillations is dropping the more layers are grown. This is because the electron beam was focused on the original surface and gets out of focus the more layers are grown. Note that the figure is only a model similar in shape to those used by film growth experts.

RHEED is an extremely popular technique for monitoring the growth of thin films. In particular, RHEED is well suited for use with molecular beam epitaxy (MBE), a process used to form high quality, ultrapure thin films under ultrahigh vacuum growth conditions. The intensities of individual spots on the RHEED pattern fluctuate in a periodic manner as a result of the relative surface coverage of the growing thin film. figure shows an example of the intensity fluctuating at a single RHEED point during MBE growth.

Each full period corresponds to formation of a single atomic layer thin film. The oscillation period is highly dependent on the material system, electron energy and incident angle, so researchers obtain empirical data to correlate the intensity oscillations and film coverage before using RHEED for monitoring film growth.

RHEED-TRAXS

Reflection high energy electron diffraction - total reflection angle X-ray spectroscopy is a technique for monitoring the chemical composition of crystals. RHEED-TRAXS analyzes X-ray spectral lines emitted from a crystal as a result of electrons from a RHEED gun colliding with the surface.

RHEED-TRAXS is preferential to X-ray microanalysis (XMA)(such as EDS and WDS) because the incidence angle of the electrons on the surface is very small, typically less than 5°. As a result, the electrons do not penetrate deeply into the crystal, meaning the X-ray emission is restricted to the top of the crystal, allowing for real-time, in-situ monitoring of surface stoichiometry.

The experimental setup is fairly simple. Electrons are fired onto a sample causing X-ray emission. These X-rays are then detected using a silicon-lithium Si-Li crystal placed behind beryllium windows, used to maintain vacuum.

MCP-RHEED

MCP-RHEED is a system in which an electron beam is amplified by a micro-channel plate (MCP). This system consists of an electron gun and an MCP equipped with a fluorescent screen opposite to the electron gun. Because of the amplification, the intensity of the electron beam can be decreased by several orders of magnitude and the damage to the samples is diminished. This method is used to observe the growth of insulator crystals such as organic films and alkali halide films, which are easily damaged by electron beams.

Reflection High-energy Electron Diffraction

Reflection high-energy electron diffraction (RHEED) is another type of diffraction technique used to characterize the surface of crystalline materials.

- RHEED systems gather information only from the surface layer of the sample and hence quite useful to monitor the growth condition of the materials at different stages:

  ☆ Substrate thermal cleaning monitoring

  ☆ Controlling initial growth stage

  ☆ Monitoring surface structure and growth dynamics

☆ Growth rate measurement

- The small incident angle makes it sensitive to the structure of top monolayers.

RHEED: Principle of Operation

RHEED is typically based on the reflection of electrons with high kinetic energy (~ 5 - 100 keV) and low impact angle (< 5°) from the surface of a solid.

- A brief illustration of the operation principle for a cubic lattice sample is described below in figure. The electrons with a momentum ko incident on the sample surface at a small incident angle. As a result, they are scattered from the top layer of atoms of the sample.

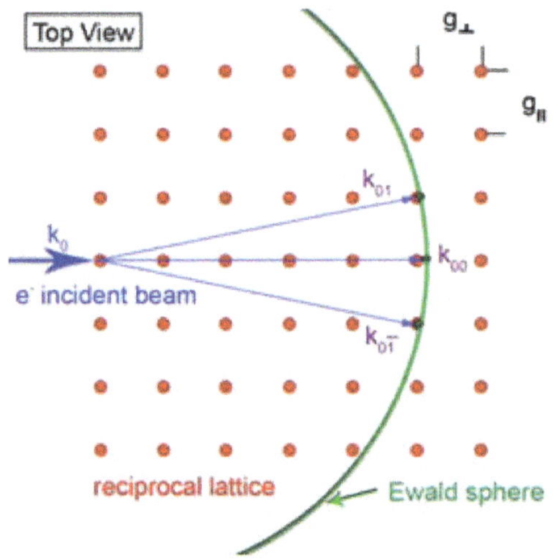

Schematic of e-beam incident on a reciprocal lattice.

In the figure, the surface is shown in reciprocal space. This reciprocal lattice builds a surface with a quadratic array of atoms. Assuming elastic scattering, the scattered wave vector kij lies on the surface of the sphere of constant energy, the so-called Ewald sphere.

- As RHEED is typically based on the reflection of electrons with high kinetic energy (~ 5 - 100 keV) and low impact angle (< 5°) from the surface of a solid, a brief illustration of the operation principle for a cubic lattice sample is described below using figure:

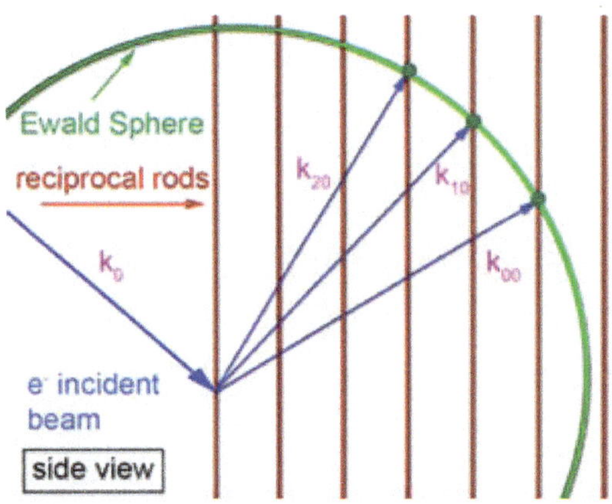

Schematic of operation principle of RHEED.

- The side view is shown in the figure. In reciprocal space, the two-dimensional array of the surface atoms turns into vertical lines, i.e., the reciprocal rods. The condition for constructive interference of the elastically scattered electron beams from the surface is fulfilled whenever these rods cross the Ewald sphere, as shown in figure.
- Figure shows the directions of elastically scattered electrons in real space. The scattered electrons hit a fluorescent screen in certain spots, lying on so-called Laue circles, which are numbered starting from zero (red) and 1 (cyan).

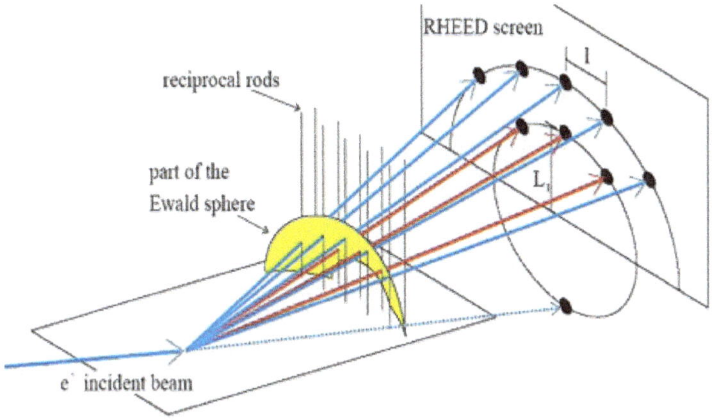

The directions of the elastically scattered electrons in real space.

The spots forming a characteristic RHEED pattern depend on the morphology and roughness on the surface. Hence, they become a powerful tool for in-situ analysis of film deposition.

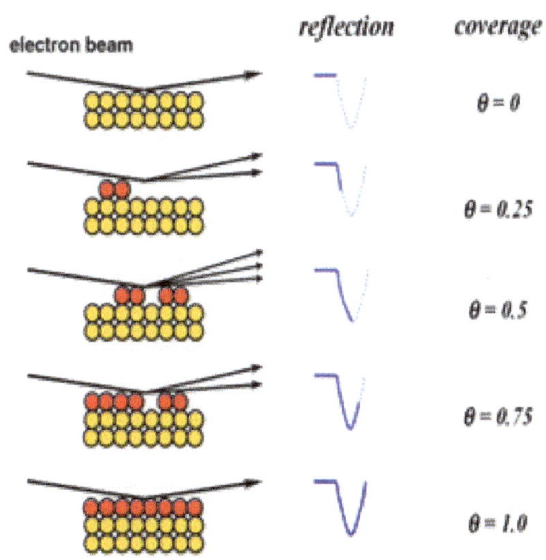

Schematic of surface coverage and corresponding reflctions in RHEED.

As the RHEED intensity depends on the film roughness, the growth process leads to characteristic intensity oscillations of the RHEED spots during the growth process with a single oscillation usually corresponding to the completion of a single monolayer, as shown in figure.

Figure demonstrates various types of RHEED patterns for different types of surfaces. The ideal smooth surface provides a clear dot patterns, which is typically distracted to be streaks for the reality smooth surfaces.

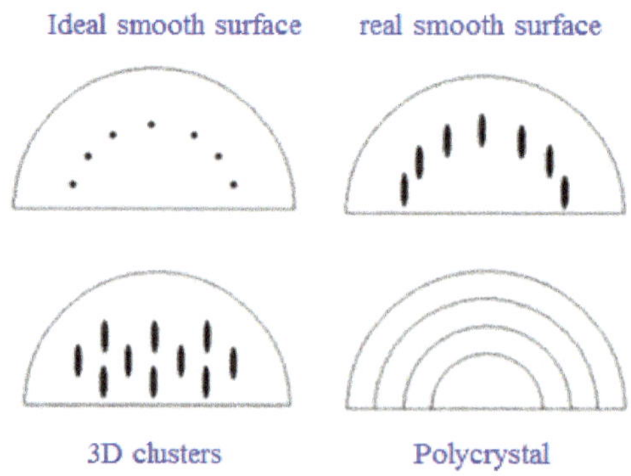

Various types of RHEED patterns for different types of surfaces.

- RHEED patterns change to different nature for typical 3D clusters and exhibit a continuous line profile for a polycrystal.

LEED versus RHEED

In LEED, low energy electrons provide large elastic scattering cross-section for back-scattered electrons with short penetration depth.

In RHEED, high energy electrons provide large elastic scattering cross-section for forward-scattered electrons with a grazing incidence to keep short penetration depth.

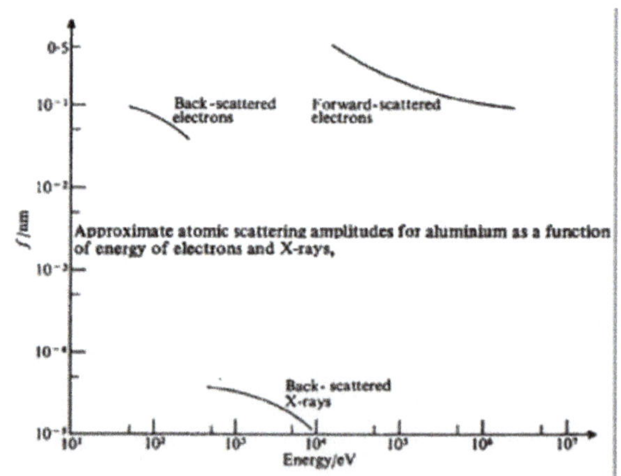

Atomic scattering amplitude for aluminium as a function of energy of electrons and X-rays.

LEED: (a) A well developed theory exists and hence quantitative analysis is possible. (b) Provides information not only in-plane but also out-of-plane. Diffraction pattern: spots. Disadvantage is the normal incidence to the sample surface.

RHEED: (a) Theory is not well developed as compared to LEED, (b) includes multiple scattering events and many more inelastic events; (c) quantitative analysis is usually not performed. Diffraction pattern: streaks; sensitive to roughness, surface must be very flat.

References

- Als-Nielsen, Jens; Mcmorrow, Des (2001). Elements of Modern X-Ray Physics. John Wiley & Sons Ltd,. pp. 40–41. ISBN 0-471-49858-0

- Brenner DJ, Hall EJ (2007). “Computed tomography- an increasing source of radiation exposure”. New England Journal of Medicine. 357 (22): 2277–2284. doi:10.1056/NEJMra072149. PMID 18046031
- Greene, Brian (1999). The Elegant Universe: Superstrings, Hidden Dimensions, and the Quest for the Ultimate Theory. New York: W.W. Norton. pp. 97–109. ISBN 0-393-04688-5
- “National Cancer Institute: Surveillance Epidemiology and End Results (SEER) data”. Seer.cancer.gov. 2010-06-30. Retrieved 2011-11-08
- Structure determination form powder diffraction data IUCr Monographs on crystallography, Edt. W.I.F. David, K. Shankland, L.B. McCusker and Ch. Baerlocher. 2002. Oxford Science publications ISBN 0-19-850091-2
- Natale, Simone (2011-08-04). “A Cosmology of Invisible Fluids: Wireless, X-Rays, and Psychical Research Around 1900”. Canadian Journal of Communication. 36 (2). ISSN 0705-3657
- Bushberg, Jerrold T.; Seibert, J. Anthony; Leidholdt, Edwin M.; Boone, John M. (2002). The essential physics of medical imaging. Lippincott Williams & Wilkins. p. 38. ISBN 978-0-683-30118-2
- “Pregnant Women and Radiation Exposure”. eMedicine Live online medical consultation. Medscape. 28 December 2008. Archived from the original on January 23, 2009. Retrieved 2009-01-16
- Glasser, Otto (1993). Wilhelm Conrad Röntgen and the early history of the roentgen rays. Norman Publishing. pp. 10–15. ISBN 0930405226
- Hiltner, PA; IM Krieger (1969). “Diffraction of Light by Ordered Suspensions”. Journal of Physical Chemistry. 73: 2306. doi:10.1021/j100727a049
- Fleming, Walter Lynwood. The South in the Building of the Nation: Biography A-J. Pelican Publishing. p. 300. ISBN 1589809467
- Schmidl, Petra G. (1996–1997). “Two Early Arabic Sources On The Magnetic Compass”. Journal of Arabic and Islamic Studies. 1: 81–132
- Attwood, David (1999). “3”. Soft X-rays and extreme ultraviolet radiation. Cambridge University Press. ISBN 978-0-521-65214-8
- Gaida, Roman; et al. (1997). “Ukrainian Physicist Contributes to the Discovery of X-Rays”. Mayo Foundation for Medical Education and Research. Archived from the original on 2008-05-28. Retrieved 2008-04-06
- Whaites, Eric; Cawson, Roderick (2002). Essentials of Dental Radiography and Radiology. Elsevier Health Sciences. pp. 15–20. ISBN 0-443-07027-X
- Karlsson, Erik B. (9 February 2000). “The Nobel Prizes in Physics 1901–2000”. Stockholm: The Nobel Foundation. Retrieved 24 November 2011
- Bushburg, Jerrold; Seibert, Anthony; Leidholdt, Edwin; Boone, John (2002). The Essential Physics of Medical Imaging. USA: Lippincott Williams & Wilkins. p. 116. ISBN 0-683-30118-7
- Markel, Howard (20 December 2012). “‘I Have Seen My Death’: How the World Discovered the X-Ray”. PBS NewsHour. PBS. Retrieved 27 April 2013
- Kasai, Nobutami; Kakudo, Masao (2005). X-ray diffraction by macromolecules. Tokyo: Kodansha. pp. 291–2. ISBN 3-540-25317-3
- “Major John Hall-Edwards”. Birmingham City Council. Archived from the original on September 28, 2012. Retrieved 2012-05-17
- Kevles, Bettyann Holtzmann (1996). Naked to the Bone Medical Imaging in the Twentieth Century. Camden, NJ: Rutgers University Press. pp. 19–22. ISBN 0-8135-2358-3
- Frame, Paul. “Wilhelm Röntgen and the Invisible Light”. Tales from the Atomic Age. Oak Ridge Associated Universities. Retrieved 2008-05-19
- Pais, Abraham (1986). Inward Bound: Of Matter and Forces in the Physical World. UK: Oxford Univ. Press. p. 79. ISBN 0-19-851997-4

Permissions

 Every chapter published in this book has been scrutinized by our experts. Their significance has been extensively debated. The topics covered herein carry significant information for a comprehensive understanding. They may even be implemented as practical applications or may be referred to as a beginning point for further studies.

We would like to thank the editorial team for lending their expertise to make the book truly unique. They have played a crucial role in the development of this book. Without their invaluable contributions this book wouldn't have been possible. They have made vital efforts to compile up to date information on the varied aspects of this subject to make this book a valuable addition to the collection of many professionals and students.

This book was conceptualized with the vision of imparting up-to-date and integrated information in this field. To ensure the same, a matchless editorial board was set up. Every individual on the board went through rigorous rounds of assessment to prove their worth. After which they invested a large part of their time researching and compiling the most relevant data for our readers.

The editorial board has been involved in producing this book since its inception. They have spent rigorous hours researching and exploring the diverse topics which have resulted in the successful publishing of this book. They have passed on their knowledge of decades through this book. To expedite this challenging task, the publisher supported the team at every step. A small team of assistant editors was also appointed to further simplify the editing procedure and attain best results for the readers.

Apart from the editorial board, the designing team has also invested a significant amount of their time in understanding the subject and creating the most relevant covers. They scrutinized every image to scout for the most suitable representation of the subject and create an appropriate cover for the book.

The publishing team has been an ardent support to the editorial, designing and production team. Their endless efforts to recruit the best for this project, has resulted in the accomplishment of this book. They are a veteran in the field of academics and their pool of knowledge is as vast as their experience in printing. Their expertise and guidance has proved useful at every step. Their uncompromising quality standards have made this book an exceptional effort. Their encouragement from time to time has been an inspiration for everyone.

The publisher and the editorial board hope that this book will prove to be a valuable piece of knowledge for students, practitioners and scholars across the globe.

Index